STANDARD FORMS

$ax^2 + bc + c = 0$ Quadratic equation in x if $a \neq 0$.

$\dfrac{y_2 - y_1}{x_2 - x_1} = m$ Slope of a straight line.

$\dfrac{y - y_1}{x - x_1} = \dfrac{y_2 - y_1}{x_2 - x_1}$ Two-point form of a straight line.

$y = mx + b$ Slope-intercept form of a straight line.

$y - y_1 = m(x - x_1)$ Point-slope form of a straight line.

$(x - h)^2 + (y - k)^2 = r^2$ Circle with center at (h, k) and radius r.

$\dfrac{(x - h)^2}{a^2} + \dfrac{(y - k)^2}{b^2} = 1$ Ellipse with center at (h, k).
 The distance from the center to the foci is c.
 If $a > b$, the ellipse is horizontal and $a^2 - c^2 = b^2$.
 If $a < b$, the ellipse is vertical and $a^2 + c^2 = b^2$.

$\dfrac{(x - h)^2}{a^2} - \dfrac{(y - k)^2}{b^2} = \pm 1$ Hyperbola with center at (h, k).
 The distance from the center to the foci is c and $a^2 + b^2 = c^2$.
 If the right side is $+1$, the hyperbola is horizontal.
 If the right side is -1, the hyperbola is vertical.

$(x - h)^2 = 4p(y - k)$ Vertical parabola.
 Vertex (h, k); focus $(h, k + p)$; directrix $y = k - p$.
 The parabola opens upward if $p > 0$ and downward if $p < 0$.

$(y - k)^2 = 4p(x - h)$ Horizontal parabola.
 Vertex (h, k); focus $(h + p, k)$; directrix $x = h - p$.
 The parabola opens to the right if $p > 0$ and to the left if $p < 0$.

College Algebra • A Skills Approach

College Algebra

A Skills Approach • LECTURE VERSION

J. Louis Nanney and John L. Cable
Miami-Dade Community College

Allyn and Bacon, Inc. • *Boston • London • Sydney • Toronto*

Series Editor: Carl Lindholm
Production Editor: Mary Hill
Designer: Dorothy Thompson
Art Editor: Armen Kojoyian
Cover Designer: Christy Rosso
Preparation Buyer: Linda Card

1980 Lecture Version of *College Algebra: A Skills Approach*, Copyright © 1978 by Allyn and Bacon, Inc., 470 Atlantic Avenue, Boston, Massachusetts 02210.

All rights reserved. No part of the material protected by this copyright notice may be reproduced or utilized in any form or by any means, electronic or mechanical, including photocopying, recording, or by any information storage and retrieval system, without written permission from the copyright owner.

Portions of this book first appeared in *Developing Skills in Algebra: A Lecture Worktext*, Second Edition, by J. Louis Nanney and John L. Cable, Copyright © 1976, 1974 by Allyn and Bacon, Inc.

Library of Congress Cataloging in Publication Data

Nanney, J   Louis.
   College algebra, a skills approach.

   Portions of this book published in the author's Developing skills in algebra, 2d ed., © 1976.
   Includes index.
   1. Algebra.   I. Cable, John Laurence, 1934- joint author.   II. Title.
QA154.2.N36 1980   512.9   79-22998
ISBN 0-205-06914-2

Cover Illustration: Four of twenty-four variations from the watercolor *Vibration Story* by Kurt Kranz. From the collection of Guido Goldman.

Printed in the United States of America.

Contents

Preface ix

CHAPTER 1 **A BRIEF REVIEW OF THE BASIC PRINCIPLES OF ALGEBRA**

1.1 The Real Numbers 2
1.2 Exponents and Radicals 7
1.3 Algebraic Expressions 16
1.4 Factoring 21
1.5 Algebraic Fractions 24
Chapter Review 27
Practice Test 28

CHAPTER 2 **EQUATIONS AND INEQUALITIES**

2.1 First-Degree Equations—One Variable 32
2.2 Equations of Higher Degree 36
2.3 Complex Numbers 42
2.4 Equations Involving Quadratic Form, Radicals, and Absolute Value 48
2.5 First-Degree Inequalities 51
2.6 Other Inequalities in One Variable 56
2.7 Verbal Problems 59
Chapter Review 63
Practice Test 64

CHAPTER 3 **RELATIONS AND FUNCTIONS**

3.1 Definitions 68
3.2 Graphs of Relations 71
3.3 The Conic Sections 79

3.4 Linear Functions 93
3.5 Some Algebra of Functions 98
3.6 Inverse Functions 100
    Chapter Review 105
    Practice Test 107

CHAPTER 4   SYSTEMS OF EQUATIONS AND INEQUALITIES

4.1 Solving Systems of Equations in Two Variables by Graphing 110
4.2 The Algebraic Solution of a System of Two Linear Equations 113
4.3 The Algebraic Solution of Three Equations with Three Variables 121
4.4 Matrices 126
4.5 Evaluating Second- and Third-Order Determinants 131
4.6 Solving Systems of First-Degree Equations by Determinants (Two Equations with Two Variables) 136
4.7 Solving Systems of First-Degree Equations by Determinants (Three Equations with Three Variables) 140
4.8 Solving Systems of Inequalities in Two Variables by Graphing 143
    Chapter Review 150
    Practice Test 152

CHAPTER 5   LOGARITHMS

5.1 Definition of Logarithm 156
5.2 Laws of Logarithms 159
5.3 Table of Logarithms 161
5.4 Antilogarithms and Interpolation 166
5.5 Multiplication and Division Using Logarithms 172
5.6 Finding Powers and Roots Using Logarithms 174
5.7 Logarithms to Different Bases 176
5.8 Exponential and Logarithmic Equations 179
    Chapter Review 183
    Practice Test 184

CHAPTER 6   POLYNOMIALS

6.1 Properties of Polynomials 188
6.2 Division of Polynomials 191
6.3 Synthetic Division 198
6.4 Zeros of Polynomials with Complex Coefficients 202
6.5 Zeros of Polynomials with Real Coefficients 205
6.6 Zeros of Polynomials with Rational or Integral Coefficients 208

      6.7   Some Theorems on the Boundaries for the Zeros of Polynomials   212
      6.8   Partial Fractions   218
            Chapter Review   223
            Practice Test   225

CHAPTER 7   SEQUENCES AND SERIES

      7.1   Mathematical Induction   228
      7.2   Sequences and Series   231
      7.3   Arithmetic Sequences   236
      7.4   Geometric Sequences   239
            Chapter Review   244
            Practice Test   245

CHAPTER 8   THE BINOMIAL THEOREM AND AN INTRODUCTION TO PROBABILITY

      8.1   Permutations   250
      8.2   Combinations   254
      8.3   The Binomial Theorem   256
      8.4   An Introduction to Probability   262
            Chapter Review   273
            Practice Test   274

Common Logarithms   276

Answers to Odd-Numbered Exercises   279

Index   305

Preface

College Algebra: A Skills Approach is a text written for the student. Explanations are complete, but as clear and concise as possible so as to keep the amount of required reading to a minimum. These explanations are followed by completely worked examples which reinforce the ideas presented.

This text is not intended to be a rigorous structural development in the sense of "modern algebra." However, as skills are developed, a greater number of proofs are given so that more rigor is achieved as the student becomes better able to appreciate and understand formal structure. Hence—"A skills approach."

Chapter 1 is a brief review of the topics generally covered in elementary and intermediate algebra. It is intended that this chapter be covered very rapidly, giving time for topics often not reached in a college algebra course. If the student finds Chapter 1 anything more than a review, then he or she probably lacks the necessary background for college algebra.

The text is structured to involve the student in "doing" the mathematics and includes many helpful features. In addition to the detailed explanations and carefully worked examples, each chapter has a chapter review and a practice test. Throughout the text, "Cautions" are given to aid the student in avoiding common errors. "Goals" at the beginning of each section help the student to know what is expected. A student study guide is also available as a companion to the text.

Although the text is designed for a one-semester or one-quarter course, if time does not permit complete coverage, certain topics can be omitted or abbreviated without destroying the continuity of the material. Some such topics are the Conic Sections, Computations with Logarithms, Sequences and Series, and Probability.

Acknowledgments

We wish to thank Carl Lindholm, mathematics editor, and Mary Hill, production editor, whose many suggestions and constant support helped to make this book possible. Also, we are indebted to the following people who reviewed the manuscript and offered many helpful suggestions:

Cal Carlson	Brainerd Community College
Mary Jane Causey	University of Mississippi
Charles Cook	Tri-State University
Vivian Dennis	Eastfield College
James B. Derr	University of West Virginia
Bruce E. Earnley	North Essex Community College
Duncan O. Faus	Monterey Peninsula College
Hank Harmeling, Jr.	North Shore Community College
Warland Hersey	North Shore Community College
Jerry Karl	Golden West College
Alan Olinsky	Bryant College
Adele Shapiro	Central Piedmont Community College
Peter Sherman	University of Oregon
John Snyder	Sinclair Community College
John Tobey	North Shore Community College
John Whitcomb	University of North Dakota
Edward Zanella	Rhode Island Junior College

College Algebra • A Skills Approach

A Brief Review of the Basic Principles of Algebra

1

A study of algebra and trigonometry requires skills from arithmetic and elementary algebra. In this text, some of these skills are assumed; others are briefly reviewed. Skills in arithmetic—adding, subtracting, multiplying, and dividing whole numbers, decimals, and fractions—are assumed. Also the ability to operate with signed numbers is assumed.

This first chapter is a brief review of topics from elementary algebra with which the student should already be familiar.*

1.1 THE REAL NUMBERS

Goals

Upon completion of this section you should be able to:

1. Recognize and use the properties of the real numbers.
2. Determine the order of any two real numbers.
3. Find the absolute value of a real number.

* *

The numbers first encountered in elementary arithmetic are those used in counting, or the set of *counting numbers*,

$$\{1, 2, 3, 4, \ldots\}$$

This set is later extended to include zero and the negatives of the counting numbers giving us the set of *integers*,

$$\{\ldots, -4, -3, -2, -1, 0, 1, 2, 3, 4, \ldots\}$$

The set of *rational numbers*—those that can be expressed as a ratio of two integers—are generally referred to as *fractions*. This set includes

---

*If the student finds this review insufficient, a careful review of an elementary algebra text such as Nanney and Cable, *Developing Skills in Algebra* (3d ed. Boston: Allyn and Bacon, Inc., 1979) is advised.

the integers: for example, 3 can be expressed as $\frac{3}{1}$, $\frac{6}{2}$, etc. In fact, this set includes all number expressions that involve only the operations of addition, subtraction, multiplication, and division.

Next, a set of numbers called the *irrational numbers*, which cannot be expressed as the ratio of integers, is developed. This set includes such numbers as π, $\sqrt{5}$, $\sqrt[3]{7}$, etc. (The decimal expansion of these numbers will neither repeat nor terminate.)

The sets of rational and irrational numbers together make up a set called the *real numbers*. Elementary algebra is a study of the real numbers and their properties.

The basic set of properties (axioms) of the real numbers can be used to justify all of the manipulations used in both arithmetic and elementary algebra. If this set of properties is accepted as being true, then all other properties can be proved as theorems. Even though an intuitive approach rather than one of rigorous proofs is used in this text, mention is often made of these basic properties. They are listed here for easy reference.

Properties of the Real Numbers

1. The real numbers are *closed* under addition. If a and b are real numbers, then $(a + b)$ is also a real number.
2. Addition of real numbers is *commutative*. If a and b are real numbers, then $a + b = b + a$.
3. Addition of real numbers is *associative*. If a, b, and c are real numbers, then $(a + b) + c = a + (b + c)$.
4. There is a real number which is the *additive identity*. This number is zero. For all real numbers a: $a + 0 = a$.
5. Each real number has a *negative* (additive inverse). If a is a real number, then there is a real number $(-a)$ such that $a + (-a) = 0$.
6. The real numbers are *closed* under multiplication. If a and b are real numbers, then ab is also a real number.
7. Multiplication of real numbers is *commutative*. If a and b are real numbers, then $ab = ba$.
8. Multiplication of real numbers is *associative*. If a, b, and c are real numbers, then $(ab)c = a(bc)$.
9. There is a real number that is the *multiplicative identity*. This number is 1. For all real numbers a: $a(1) = a$.
10. Each nonzero real number has a *reciprocal* (multiplicative inverse). If a is a real number (not zero), then there is a real number $\frac{1}{a}$ such

1.1 THE REAL NUMBERS

that $a\left(\dfrac{1}{a}\right) = 1$.

11. The real numbers obey the *distributive property* of multiplication over addition. If a, b, and c are real numbers, then $a(b + c) = ab + ac$.

Any set of elements in which all eleven of these properties are true is called a *field*. The field of real numbers is a subset of another field called the *complex numbers* which will be studied in Chapter 2.

The field of real numbers has another property called the *ordering property*.

Ordering Property

There is a subset of the set of real numbers (i.e., a set of numbers belonging to the real numbers) called the *positive* numbers such that:

1. This set (the positive real numbers) is closed under the operations of addition and multiplication.
2. For every real number a, one and only one of the following statements is true:
   a. a is a member of the set of positive real numbers.
   b. a is the negative of a member of the set of positive real numbers.
   c. a is zero.

Since all real numbers are classified as positive, negative, or zero by the ordering property, this set can be represented on a number line. By using a straight line from plane geometry (a straight line has infinite length) and choosing a point to be represented by 0 and another point to be represented by 1, we can, with certain agreements, establish a correspondence between the real numbers and the points of the line. We first agree to place the positive numbers to the right of 0, and the negative numbers to the left of 0. We also agree that the length of the line segment from 0 to 1 will be used as a unit measurement between all real numbers that differ by 1 and that real numbers between 0 and 1 will be represented by proportional parts of this unit. An illustration of such a number line follows.

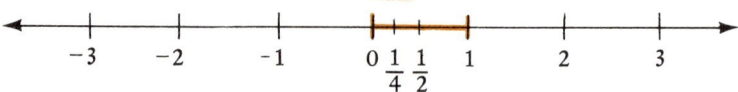

The number line is a very useful tool in the study of mathematics. One example of its use is in the study of inequalities. The symbol for inequality "<" can be defined in terms of the number line.

Definition   If a and b are real numbers, $a < b$ means that a is to the left of b on the number line.

This same symbol is sometimes turned in the other direction ">." $a < b$ is usually read "a is less than b," and $b > a$ is usually read "b is greater than a." Of course, both expressions mean the same thing and could be read in either direction. A good way to remember the meaning of the symbol is to recognize that the pointed end is always in the direction of the smaller number (i.e., the number which would be placed to the left of the other on the number line).

Exercise 1.1.1

Place the proper symbol, $<$ or $>$, for the question mark in each of the following to make a true statement.

1. 4 ? 6
2. 9 ? 2
3. −3 ? 0
4. 5 ? 0
5. −1 ? −5
6. 3 ? −6
7. $\frac{3}{2}$? 2
8. $-\frac{10}{3}$? −3
9. −0.5 ? −0.6
10. $\frac{1}{3}$? 0.33

Distance, without regard to direction, is defined by the term *absolute value*. The absolute value of an expression is indicated by placing vertical bars before and after the expression. $|a - b|$ is read as "the absolute value of $a - b$." $|x|$ is read as "the absolute value of x." $|a - b|$ implies the distance from a to b or the distance from b to a, without regard to direction. A formal definition follows.

Definition
$$|x| = \begin{cases} x & \text{if } x \text{ is positive} \\ 0 & \text{if } x \text{ is zero} \\ -x & \text{if } x \text{ is negative} \end{cases}$$

An immediate result of this definition is that the absolute value of an expression is always nonnegative. If x is positive or zero, it is clear from the definition that $|x|$ is nonnegative. Also if x is negative, then $|x| = -x$ (opposite of x) and is positive.

Absolute value is used in mathematics to measure the "nearness" of points.

Exercise 1.1.2

1. Evaluate the following:

   a. $|9 - 3|$
   b. $|5 - 6|$
   c. $|3 - 3|$
   d. $|10 - 7|$
   e. $|7 - 10|$
   f. $|8 - (-5)|$
   g. $\left|\dfrac{2}{5} - \dfrac{3}{8}\right|$
   h. $\left|\dfrac{3}{7} - \dfrac{4}{9}\right|$

2. Justify each of the following by stating *one* of the field properties:

   a. $4 + (2 + 3) = (4 + 2) + 3$
   b. $5 + 0 = 5$
   c. $(6)(1) = 6$
   d. $(4)\left(\dfrac{1}{4}\right) = 1$
   e. $5(2 + 3) = 5(2) + 5(3)$
   f. $8 + (-8) = 0$
   g. $(2 + 3) + 6 = (3 + 2) + 6$
   h. $[(4)(2)](3) = (3)[(4)(2)]$

3. Determine which of the following sets are closed under (*a*) addition and (*b*) multiplication:

   a. { integers }
   b. { rational numbers }
   c. { all positive real numbers }
   d. { all negative real numbers }
   e. { even counting numbers }
   f. { odd counting numbers }
   g. { 0 }
   h. { -1, 0, 1 }

4. The value of π to five decimal places is 3.14159. A less accurate value of $\dfrac{22}{7}$ is often used for π in elementary work. Find the decimal approximation of $\dfrac{22}{7}$ and then substitute $<$ or $>$ for the question mark in the following to make a true statement:

$$\dfrac{22}{7} \;?\; \pi$$

5. Restate the three statements in part 2 of the ordering property by using the symbols $>$, $<$, and $=$.

6. a. What is the largest integer less than or equal to π?
   b. What is the largest rational number less than or equal to π?
   c. What is the largest real number less than or equal to π?

1.2 EXPONENTS AND RADICALS

Goals

Upon completion of this section you should be able to:

1. Define positive whole number exponents, the zero exponent, negative exponents, and fractional exponents.
2. Use the laws of exponents and the definitions to simplify expressions containing exponents.
3. Define a radical and simplify expressions containing radicals.

* *

Definition

$x^n = \underbrace{(x)(x)(x)(x) \cdots (x)}_{n \text{ factors}}$, if n is a positive integer.

In words: "x to the nth power, when n is a positive integer, implies that x is used as a factor n times."

This definition gives rise to the following laws of exponents.

Law I

$x^a \cdot x^b = x^{a+b}$

To multiply like bases, add the exponents.

Example $\qquad x^3 \cdot x^5 = x^8$

Law II

$x^a \div x^b = x^{a-b} \quad x \neq 0$

To divide like nonzero bases, subtract the exponent of the divisor from the exponent of the dividend.

Example $\qquad x^7 \div x^5 = x^2$

Law III

$(x^a)^b = x^{ab}$

To raise a power to a power, multiply exponents.

Example $\qquad (x^3)^5 = x^{15}$

Law IV

$(xy)^a = x^a y^a$

The power of a product is the product of the powers.

Example $$(xy)^3 = x^3y^3$$

Law V $$\left(\frac{x}{y}\right)^a = \frac{x^a}{y^a} \quad y \neq 0$$

The power of a quotient is the quotient of the powers.

Example $$\left(\frac{x}{y}\right)^4 = \frac{x^4}{y^4}$$

Law II of exponents gives rise to the necessity of two further definitions to expand the meaning of exponents. As written in Law II, $x^a \div x^b = x^{a-b}$, $x \neq 0$, actually would apply only when $a > b$, if we restrict ourselves to positive integers. To expand the concept of exponent, let us consider the cases when $a = b$ and $a < b$.

If $a = b$ and $x \neq 0$, then

$$\frac{x^a}{x^a} = x^{a-a} = x^0$$

but

$$\frac{x^a}{x^a} = 1$$

since any nonzero quantity divided by itself is 1. This observation gives rise to the following definition.

Definition $x^0 = 1$, for all $x \neq 0$.

When $a < b$, $\frac{x^a}{x^b} = x^{a-b}$ would give rise to a negative exponent. For instance,

$$\frac{x^3}{x^5} = x^{3-5} = x^{-2}$$

From the definition of exponent, however, and a basic principle of simplifying fractions, we have

$$\frac{x^3}{x^5} = \frac{(x)(x)(x)}{(x)(x)(x)(x)(x)} = \frac{1}{x^2}$$

Hence the following definition.

Definition $x^{-a} = \frac{1}{x^a}$, a is a positive integer and $x \neq 0$.

A BRIEF REVIEW OF THE BASIC PRINCIPLES OF ALGEBRA

Notice that the definitions are consistent with the laws of exponents, and therefore Laws I to V will still be valid when we allow exponents to be *any* integer.

Exercise 1.2.1

Simplify the following. Leave all answers with only positive exponents. Assume all expressions represent real numbers.

1. $(2x^2)^4$
2. $(3xy^2)^3$
3. $\left(\dfrac{2}{x^2 y}\right)^3$
4. $[(x^2)^3]^4$
5. $\left(\dfrac{xy^2}{x^2 y}\right)^2$
6. $\left(\dfrac{2x^3}{xy^2}\right)^3$
7. $[2(xy)^2]^3$
8. $\left(\dfrac{1}{3x^2}\right)^2$
9. $\dfrac{(x^2)^3}{(x^3)^2}$
10. $\dfrac{(-3x)^3}{9x}$
11. $\dfrac{-4x^2}{(2x)^3}$
12. $\dfrac{(2x^2 y^3)^3}{(2x^3 y^2)^2}$
13. $2x^2 y (x^2 y)^3$
14. $(-x^2 y)^3 (-2xy^3)^2$
15. $[2(xy)^2]^5 [3(x^2 y)^3]^2$
16. $\left(\dfrac{x}{y}\right)^3 \left(\dfrac{2x}{y}\right)^4$
17. $\left(\dfrac{-2}{x}\right)^3 \left(\dfrac{x}{2}\right)^2$
18. $\left(\dfrac{x^2 y}{3}\right) \left(\dfrac{3}{xy^2}\right)^2$
19. $\left(\dfrac{-2x}{5y^2}\right)^3 \left(\dfrac{5y}{4x^2}\right)^2$
20. $\left(\dfrac{x^3}{8}\right)^2 \left(\dfrac{4}{x^2}\right)^3$
21. $x^{-3} y^{-5}$
22. $\left(\dfrac{a}{b}\right)^{-5}$
23. $\dfrac{1}{3^{-2}}$
24. $(ab)^{-1}$
25. $(a+b)^{-1}$
26. $a^{-1} + b^{-1}$
27. $x^{-2} y^4 z^{-1}$
28. $(x^3)^{-2}$
29. $(x^{-2} y^4)^{-4}$
30. $\dfrac{x^{-3}}{x^5}$
31. $\dfrac{x^{-2} x^{-6}}{x^7}$
32. $\dfrac{x^{-3} y^2}{x^{-5} y^{-1}}$

1.2 EXPONENTS AND RADICALS

9

33. $\dfrac{x^3 y^{-7}}{x^3 y^0}$
34. $(x^2)^{-5}(x^{-3})^2$

35. $(-2x^3)^{-3}(3x^{-1})^2$
36. $(3^{-1}x^{-4})^{-3}$

37. $\left(\dfrac{x^{-1}y^2}{y^{-3}}\right)^{-5}$
38. $(2x^2 y^{-3})^{-3}(-2x^{-3}y^2)^2$

39. $\left(\dfrac{x}{y}\right)^{-3}\left(\dfrac{x^2}{y^{-1}}\right)^0$
40. $\left(\dfrac{x^2 y^{-1}}{x^{-5}}\right)\left(\dfrac{x^{-1}y^3}{x^5}\right)^{-3}$

The symbol $\sqrt{}$ is called the *radical sign*. An expression containing the radical sign is called a *radical*. The radical $\sqrt[5]{32}$ is read "the principal fifth root of 32." The number under the radical sign (32 in this case) is called the *radicand* and the integer 5 is the *index*. If the index is 2, it is usually omitted. Hence $\sqrt{8}$ and $\sqrt[2]{8}$ each indicate the principal square root of 8. If the radical sign is used to indicate any root other than the square root, the index must be written. The index number is always a positive integer greater than 1.

EXAMPLES

$\sqrt[4]{24}$ means the principal fourth root of 24.

$\sqrt[3]{24}$ means the principal cube root of 24.

$\sqrt{24}$ means the principal square root of 24.

Definition   If $x^n = y$, then x is an nth root of y.

EXAMPLES

Since $3^2 = 9$, 3 is a square root of 9.

Since $2^4 = 16$, 2 is a fourth root of 16.

Since $(-3)^2 = 9$, -3 is a square root of 9.

Since $(-2)^4 = 16$, -2 is a fourth root of 16.

Notice from these examples that 9 has two square roots, 3 and -3. It can be shown by more advanced methods that every number has two square roots, three cube roots, four fourth roots, etc. Not all of these roots, however, are in the set of real numbers.

We use the radical sign to indicate the *principal* root of a number.

Definition $\sqrt[n]{x}$, the principal nth root of x, is

a. Zero if $x = 0$.
b. Positive if $x > 0$.
c. Negative if $x < 0$ and n is an odd integer.
d. Not a real number if $x < 0$ and n is an even integer.

EXAMPLES
$\sqrt{16} = 4$
$\sqrt[3]{8} = 2$
$\sqrt[3]{-8} = -2$
$\sqrt[4]{-16}$ is not a real number

CAUTION: The radical symbol is always the *principal* root. $x^2 = 4$ and $x = \sqrt{4}$ are not the same. If $x^2 = 4$, then $x = 2$ or $x = -2$. But if $x = \sqrt{4}$, then the only solution is $x = 2$.

Definition $x^{\frac{a}{b}} = \sqrt[b]{x^a}$

This definition expands the meaning of exponent to include all rational exponents. We expect this new definition to be consistent with the laws of exponents. For example, this definition gives

$$x^{\frac{1}{2}} = \sqrt{x}$$

and since

$$x^{\frac{1}{2}} x^{\frac{1}{2}} = x^{\frac{1}{2} + \frac{1}{2}} = x$$

by Law I, we see that the definition is consistent in this instance.

EXAMPLES
$$x^{\frac{2}{3}} = \sqrt[3]{x^2}$$

$$2^{\frac{1}{2}} = \sqrt{2}$$

$$x^{-\frac{2}{3}} = \sqrt[3]{x^{-2}} = \sqrt[3]{\frac{1}{x^2}} = \frac{1}{\sqrt[3]{x^2}}$$

1.2 EXPONENTS AND RADICALS

Radicals are in simplest form when

1. The index number cannot be smaller.
2. No factor of the radicand is raised to a power equal to or greater than the index number.

EXAMPLE

Simplify $\sqrt{75}$.

The question is: "Does 75 have a factor which is raised to the second or greater power?" To answer this question we need only to look for factors of 75 which are perfect squares.

$$\sqrt{75} = \sqrt{(25)(3)}$$
$$= \sqrt{25}\sqrt{3}$$
$$= 5\sqrt{3}$$

Note: To see that $\sqrt{(25)(3)} = \sqrt{25}\sqrt{3}$ we need only use the fourth law of exponents obtaining

$$[(25)(3)]^{\frac{1}{2}} = (25)^{\frac{1}{2}} (3)^{\frac{1}{2}}$$

EXAMPLE

Simplify $\sqrt{8x^3y^4}$.

$$\sqrt{8x^3y^4} = \sqrt{(4)(2)(x^2)(x)(y^4)}$$
$$= \sqrt{4}\sqrt{2}\sqrt{x^2}\sqrt{x}\sqrt{y^4}$$
$$= 2xy^2\sqrt{2x}$$

EXAMPLE

Simplify $\sqrt[3]{16x^4y^5}$.

$$\sqrt[3]{16x^4y^5} = \sqrt[3]{(8)(2)(x^3)(x)(y^3)(y^2)}$$
$$= 2xy\sqrt[3]{2xy^2}$$

EXAMPLE

Simplify $\sqrt[6]{x^2y^4}$ where x and y are nonnegative reals.

Step 1

$$\sqrt[6]{x^2y^4} = (x^2y^4)^{\frac{1}{6}} = x^{\frac{2}{6}} y^{\frac{4}{6}}$$

The fractional exponents are then reduced to simplest form.

Step 2
$$x^{\frac{2}{6}} y^{\frac{4}{6}} = x^{\frac{1}{3}} y^{\frac{2}{3}}$$

This step is possible only because the variables are nonnegative. If, for example, the value of x were (-8), we could not reduce the fractional exponent since

$$(-8)^{\frac{2}{6}} \neq (-8)^{\frac{1}{3}}$$

Step 3 If the fractional exponents do not have a common denominator, they are changed to a common denominator. In this case they already have a common denominator.

$$x^{\frac{1}{3}} y^{\frac{2}{3}} = (xy^2)^{\frac{1}{3}}$$

Step 4 Now we change back to radical form.

$$(xy^2)^{\frac{1}{3}} = \sqrt[3]{xy^2}$$

Therefore $\quad \sqrt[6]{x^2 y^4} = \sqrt[3]{xy^2}$

Sometimes to simplify an expression we need to remove factors as well as make the index smaller.

EXAMPLE Simplify $\sqrt[6]{25x^4 y^8}$.

$$\sqrt[6]{25x^4 y^8} = \sqrt[6]{5^2 x^4 y^8}$$
$$= 5^{\frac{2}{6}} x^{\frac{4}{6}} y^{\frac{8}{6}}$$
$$= 5^{\frac{1}{3}} x^{\frac{2}{3}} y^{\frac{4}{3}}$$
$$= \sqrt[3]{5x^2 y^4}$$

We see that we still have a factor, y, raised to a power higher than the index number 3. We further simplify

$$\sqrt[3]{5x^2 y^4} = \sqrt[3]{5x^2 y^3 y}$$
$$= y \sqrt[3]{5x^2 y}$$

The laws of exponents allow us to perform operations on radicals. For instance, if a and b are not both negative, then, by Law IV

$$\sqrt{a} \sqrt{b} = a^{\frac{1}{2}} b^{\frac{1}{2}} = (ab)^{\frac{1}{2}} = \sqrt{ab}$$

1.2 EXPONENTS AND RADICALS

And, by the distributive property and Law IV,
$$\sqrt{6}(\sqrt{5}+\sqrt{7}) = \sqrt{30} + \sqrt{42}$$
Also by the distributive property,
$$3\sqrt{5} + 2\sqrt{5} = (3+2)\sqrt{5} = 5\sqrt{5}$$

EXAMPLE Simplify $3\sqrt{2} + \sqrt{50} + \sqrt{32}$.

We first simplify $\sqrt{50}$ and $\sqrt{32}$ obtaining
$$\begin{aligned}3\sqrt{2} + \sqrt{50} + \sqrt{32} &= 3\sqrt{2} + 5\sqrt{2} + 4\sqrt{2} \\ &= (3+5+4)\sqrt{2} \\ &= 12\sqrt{2}\end{aligned}$$

EXAMPLE Simplify $\sqrt[3]{a^4} + \sqrt[3]{8a^4} - \sqrt[6]{4a^2}$.

Again, we first simplify the radicals.
$$\begin{aligned}\sqrt[3]{a^4} + \sqrt[3]{8a^4} - \sqrt[6]{4a^2} &= a\sqrt[3]{a} + 2a\sqrt[3]{a} - \sqrt[3]{2a} \\ &= 3a\sqrt[3]{a} - \sqrt[3]{2a}\end{aligned}$$

An algebraic expression containing radicals is in simplest form when
1. Each radical in the expression is in simplest form.
2. No radical appears in the denominator of a fraction.

EXAMPLE Simplify $\dfrac{3}{\sqrt{8}}$.

Here the requirement is to find a number which when multiplied by $\sqrt{8}$ yields the square root of a perfect square. Since $(\sqrt{8})(\sqrt{2}) = \sqrt{16}$, the desired number is $\sqrt{2}$. Hence,
$$\begin{aligned}\frac{3}{\sqrt{8}} &= \frac{3}{\sqrt{8}} \cdot \frac{\sqrt{2}}{\sqrt{2}} \\ &= \frac{3\sqrt{2}}{\sqrt{16}} \\ &= \frac{3\sqrt{2}}{4}\end{aligned}$$

EXAMPLE

Simplify $\dfrac{2}{\sqrt[3]{5}}$.

$$\dfrac{2}{\sqrt[3]{5}} = \dfrac{2}{\sqrt[3]{5}} \cdot \dfrac{\sqrt[3]{25}}{\sqrt[3]{25}}$$

$$= \dfrac{2\sqrt[3]{25}}{\sqrt[3]{125}}$$

$$= \dfrac{2\sqrt[3]{25}}{5}$$

EXAMPLE

Simplify $\dfrac{a-b}{\sqrt{a+b}}$.

$$\dfrac{a-b}{\sqrt{a+b}} = \dfrac{a-b}{\sqrt{a+b}} \cdot \dfrac{\sqrt{a+b}}{\sqrt{a+b}}$$

$$= \dfrac{(a-b)\sqrt{a+b}}{\sqrt{(a+b)^2}}$$

$$= \dfrac{(a-b)\sqrt{a+b}}{a+b}$$

EXAMPLE

Simplify $\dfrac{3}{\sqrt{5}+1}$.

To rationalize the denominator of this fraction, we must multiply numerator and denominator by $(\sqrt{5}-1)$.

$$\dfrac{3}{\sqrt{5}+1} = \dfrac{3}{\sqrt{5}+1} \cdot \dfrac{\sqrt{5}-1}{\sqrt{5}-1}$$

$$= \dfrac{3(\sqrt{5}-1)}{\sqrt{25}-1}$$

$$= \dfrac{3\sqrt{5}-3}{4}$$

Exercise 1.2.2

Simplify the following. Assume values of x and y that yield real numbers.

1. $\sqrt{144}$
2. $\sqrt[4]{16}$
3. $\sqrt[5]{-32}$
4. $(-27)^{\frac{2}{3}}$

1.2 EXPONENTS AND RADICALS

5. $(\sqrt[5]{-1})^3$
6. $\sqrt{27}$
7. $\sqrt{125}$
8. $\sqrt[3]{16}$
9. $\sqrt[3]{x^6 y^9}$
10. $\sqrt[8]{x^2 y^4}$
11. $\sqrt[12]{x^{10} y^6}$
12. $\sqrt[8]{4x^4}$
13. $\sqrt[3]{64 x^4 y^6}$
14. $\sqrt{144 x^{10} y^5}$
15. $\sqrt[6]{25 x^2 y^{14}}$
16. $\sqrt{3} + 4\sqrt{3} - 2\sqrt{3}$
17. $3\sqrt{5} - 2\sqrt{3} + 4\sqrt{5} - \sqrt{3}$
18. $2\sqrt{2} - 3\sqrt[3]{2} - \sqrt{8}$
19. $\sqrt{9} + \sqrt{18} - \sqrt{36}$
20. $\sqrt[3]{16} - \sqrt[3]{54} + \sqrt[3]{3}$
21. $\sqrt{50} + \sqrt[3]{2} - \sqrt{32} - \sqrt[3]{16}$
22. $\sqrt{2}\sqrt{5}$
23. $(2\sqrt{7})(3\sqrt{2})$
24. $3\sqrt{3}(2\sqrt{5} - 3\sqrt{7})$
25. $5\sqrt{3}(2\sqrt{6} + \sqrt{15})$
26. $\sqrt[3]{2}(4\sqrt[3]{4} - 2\sqrt[3]{32})$
27. $\sqrt{3}(\sqrt{2} + \sqrt{7}) + \sqrt{2}(\sqrt{5} - \sqrt{3})$
28. $(\sqrt{3} + \sqrt{2})(\sqrt{2} + \sqrt{5})$
29. $(2\sqrt{2} - 5\sqrt{7})(2\sqrt{2} + 5\sqrt{7})$
30. $\dfrac{1}{\sqrt{5}}$
31. $\dfrac{3}{\sqrt{20}}$
32. $\dfrac{3}{\sqrt[3]{2x}}$
33. $\dfrac{2y}{\sqrt[3]{4x^2 y}}$
34. $\dfrac{1}{\sqrt{x+3}}$
35. $\dfrac{8}{\sqrt{x-2}}$
36. $\dfrac{1}{\sqrt{3}-2}$
37. $\dfrac{4}{3-\sqrt{5}}$
38. $\dfrac{1}{\sqrt{3}-\sqrt{5}}$
39. $\dfrac{a-1}{\sqrt{a}-1}$
40. $\dfrac{\sqrt{3}+\sqrt{2}}{\sqrt{3}-\sqrt{2}}$

1.3 ALGEBRAIC EXPRESSIONS

Goals

Upon completion of this section you should be able to:
1. Simplify algebraic expressions by combining like terms and removing parentheses.
2. Identify polynomials, monomials, binomials, and trinomials.

* *

Algebra may be considered as a mathematical system that generalizes the operations on a set of numbers. In the process of this generalization, letters are used to represent numbers. When letters are used to represent numbers, they are called *variables*.

Definition

An *algebraic expression* is an indicated finite series of the operations of addition, subtraction, multiplication, division, and the extraction of roots on a set of numbers, or variables representing numbers from that set. (The set of numbers used in this text will be the set of real numbers unless otherwise specified.)

An example of an algebraic expression is

$$\frac{3x^2 \sqrt{y} + 1}{x + y}$$

If specific numbers are substituted for the variables in an algebraic expression, the resulting number is called the *value* of the expression.

EXAMPLE

Evaluate $\dfrac{3x^2 \sqrt{y} + 1}{x + y}$ if $x = -4$ and $y = 9$.

$$\frac{3x^2 \sqrt{y} + 1}{x + y} = \frac{3(-4)^2 \sqrt{9} + 1}{-4 + 9}$$

$$= \frac{3(16)(3) + 1}{-4 + 9}$$

$$= \frac{145}{5}$$

$$= 29$$

The following set of exercises is designed to review your skills in arithmetic and signed numbers.

Exercise 1.3.1

Evaluate the following:

1. $5x^2$, $x = 2$
2. $(5x)^2$, $x = 2$
3. $-x^2$, $x = 5$
4. $(-x)^2$, $x = 5$

5. $\dfrac{3x\sqrt{y}+4}{3}$, $x=2, y=4$ 6. $\dfrac{x^2 y - \sqrt{x}}{x}$, $x=4, y=3$

7. $\dfrac{-3x^3+\sqrt{y}}{2y}$, $x=-2, y=9$ 8. $\dfrac{x^5 y^2 - 2x^3}{\sqrt[3]{x}}$, $x=-1, y=3$

9. $(-2x)^3 \sqrt[3]{y} + \sqrt[5]{y^2}$, $x=2, y=-1$ 10. $\dfrac{x^{\frac{1}{2}} y^5 - 4xy^2}{y^{\frac{1}{2}}}$, $x=16, y=4$

In an expression that is an indicated product, the numbers (or variables) being multiplied are called *factors.*

An expression composed entirely of factors is called a *term.*

An indicated sum may be composed of many terms, such as $3x + 2yzw$.

EXAMPLES

$3xyz$ has four *factors*: 3, x, y, and z.

$2x + 3y - 5z$ has three *terms*: $2x$, $3y$, and $-5z$.

$5x(2x - 3) + 1$ has two *terms*: $5x(2x - 3)$ and 1.

Certain algebraic expressions that are commonly used are given special names.

Definition

An algebraic expression in which the only operation is multiplication of numbers and nonnegative integral powers of a variable is called a *monomial.*

Definition

An algebraic expression that is the sum of one or more monomials with respect to the same variable is called a *polynomial* in that variable. Note that a monomial is a polynomial.

EXAMPLES

$3x^2 y$ is a monomial with respect to x and is also a monomial with respect to y.

$7x\sqrt{y}$ is a monomial with respect to x but *not* a monomial with respect to y. Why?

$3x^2 y + 7x\sqrt{y}$ is a polynomial with respect to x.

$3x^2 - 2x + 4$ is a polynomial with respect to x. (We can think of the term 4 as $4x^0$.)

18 A BRIEF REVIEW OF THE BASIC PRINCIPLES OF ALGEBRA

Polynomials with two terms are called *binomials* and with three terms are called *trinomials*.

Exercise 1.3.2

For each of the following, answer "yes" or "no" to the questions *a* to *d*:

a. A monomial with respect to *x*?
b. A monomial with respect to *y*?
c. A polynomial with respect to *x*?
d. A polynomial with respect to *y*?

1. $5x^2y$
2. $3xy^5 + 4x^2y$
3. $4x^{\frac{1}{2}} + xy^3$
4. $3x^4 + 5y$
5. $x\sqrt{5}$
6. $5\sqrt{x} + \dfrac{y\sqrt{6}}{2}$
7. $3x\sqrt{y} - 5y$
8. $2x^2y^3 - 6xy^8 + \dfrac{1}{y}$

In an expression such as $3xy$, 3 is called the *coefficient* and *x* and *y* are called the *literal factors*.

Definition Two monomials, or terms of a polynomial, are *like terms* if and only if the literal factors of one are identical to the literal factors of the other.

This definition is important because *only* like terms can be combined.

Rule To combine like terms, combine the numerical coefficients and use this result as the coefficient of the common literal factors of the terms. (This is an application of the distributive property.)

EXAMPLES
$3x^2y - 7x^2y + 10xy = -4x^2y + 10xy$
$2(x + y) + 5(x + y) = 7(x + y)$
$2x + 3y + 3x - 7xy + 2y - 7 = 5x + 5y - 7xy - 7$

To find the product of a monomial and a polynomial having two or more terms, we need simply to apply the distributive property.

1.3 ALGEBRAIC EXPRESSIONS

$$3ab(3x + 2y) = 9abx + 6aby$$

The product of two polynomials is also found by using the distributive property. We first regard one of the polynomials as a monomial.

$$(3x + 2y)(a + b + c) = (3x + 2y)a + (3x + 2y)b + (3x + 2y)c$$

We use the distributive property again to complete the process giving

$$3ax + 2ay + 3bx + 2by + 3cx + 2cy$$

A polynomial is simplified if all like terms are combined.

EXAMPLES

$$3x^2y(2x + 5y) = 6x^3y + 15x^2y^2$$

$$\begin{aligned}(x + 2y - z)(2x - 3y + z) &= 2x(x + 2y - z) - 3y(x + 2y - z) \\ &\quad + z(x + 2y - z) \\ &= 2x^2 + 4xy - 2xz - 3xy - 6y^2 \\ &\quad + 3yz + xz + 2yz - z^2 \\ &= 2x^2 + xy - xz + 5yz - 6y^2 - z^2\end{aligned}$$

A pattern develops when we multiply a binomial by a binomial that is very useful both in multiplying and in factoring.

$$(a + b)(c + d) = ac + ad + bc + bd$$

Polynomials will be discussed in more detail in Chapter 6.

Exercise 1.3.3

Simplify the following:

1. $3x^2y - 2xy^2 + 5xy^2 - x^2y$
2. $5ab - 4a^2b + 6ab - 9a^2b$
3. $2x(x + y) - 3y(x - 2y)$
4. $4x(x + 3) - (5x - 2)$
5. $x(a + b) - (a + b)$
6. $(x + 3)(x - 2)$
7. $(x + 3)(x^2 - 2x + 3)$
8. $(x + 5)(x - 5)$
9. $(x + 3)(x^2 - 3x + 9)$
10. $(2x - 5)(4x^2 + 10x + 25)$
11. $(a + b)(a - b)$
12. $(a - b)(a^2 + ab + b^2)$
13. $(a + b)(a^2 - ab + b^2)$

1.4 FACTORING

Goals

Upon completion of this section you should be able to:

1. Factor an algebraic expression or determine that it is prime.

* *

Factoring an algebraic expression is that process which takes a sum of terms and changes it into an equivalent product of factors. In one sense it is the opposite of multiplication, and multiplication of the factors to give the original expression is always a check as to the correctness of the factoring. To be completely factored, each factor must be prime, and the product of the factors must give the original expression.

Problems in factoring usually fall into one or more of the following categories:

1. Removing the greatest common factor.
2. The difference of two perfect squares.
3. The sum or difference of two perfect cubes.
4. The general trinomial.
5. Factoring by grouping the terms so they fall into one of the above categories.

A problem in the first category merely involves the use of the distributive property.

EXAMPLE

Factor $3x^2y + 6xyz - 18xy^2z$.

We remove the greatest common factor $3xy$ giving
$$3xy(x + 2z - 6yz)$$

To factor problems in the second and third categories, the following formulas should be memorized:

1. $a^2 - b^2 = (a + b)(a - b)$
2. $a^3 - b^3 = (a - b)(a^2 + ab + b^2)$
3. $a^3 + b^3 = (a + b)(a^2 - ab + b^2)$

EXAMPLE

Factor $8x^3 + 27y^3$.
$$8x^3 + 27y^3 = (2x)^3 + (3y)^3$$

so we can use formula (3) getting
$$8x^3 + 27y^3 = (2x + 3y)(4x^2 - 6xy + 9y^2)$$

A problem in the fourth category, a general trinomial such as
$$4x^2 + 15xy + 9y^2$$
can be factored by "trial and error" or by other methods explained in various elementary algebra texts.
$$4x^2 + 15xy + 9y^2 = (4x + 3y)(x + 3y)$$
This answer can be checked by using the pattern for multiplying two binomials developed in the previous section.

We use the fifth category, factoring by grouping, when there are four or more terms in an expression. The process of grouping is generally for the purpose of identifying the problem as one or more of the types just discussed.

EXAMPLE

The polynomial $4x^2 + y^2 - 9 - 4xy$ when grouped as
$$(4x^2 - 4xy + y^2) - 9$$
or
$$(2x - y)^2 - 3^2$$
is recognized as the difference of two perfect squares and then can be factored as
$$(2x - y + 3)(2x - y - 3)$$

EXAMPLES

Factor completely $16x^2 - 20x - 6$.
$$16x^2 - 20x - 6 = 2(8x^2 - 10x - 3)$$
$$= 2(4x + 1)(2x - 3)$$

Factor completely $xy + 2 + y + 2x$. Grouping gives
$$xy + 2 + y + 2x = (xy + y) + (2x + 2)$$
$$= y(x + 1) + 2(x + 1)$$
$$= (x + 1)(y + 2)$$

Factor completely $54x^3 - 16$.
$$54x^3 - 16 = 2(27x^3 - 8)$$

$$= 2[(3x)^3 - (2)^3]$$
$$= 2(3x - 2)(9x^2 + 6x + 4)$$

Factor completely $4x^2 - 25y^2 - 10y - 12x + 8$. If we arrange this expression as

$$4x^2 - 12x - 25y^2 - 10y + 8$$

and notice that 8 could be written as $9 - 1$, we can obtain

$$(4x^2 - 12x + 9) - (25y^2 + 10y + 1)$$

or
$$(2x - 3)^2 - (5y + 1)^2$$

which is the difference between two perfect squares giving

$$[(2x - 3) - (5y + 1)] \, [(2x - 3) + (5y + 1)]$$

or
$$(2x - 5y - 4)(2x + 5y - 2)$$

Factor completely $5(x + y)^2 + 7(x + y) + 2$. This expression is in the form of a general trinomial. If we regard $(x + y)$ as the variable and factor, we obtain

$$[5(x + y) + 2] \, [(x + y) + 1]$$

or
$$(5x + 5y + 2)(x + y + 1)$$

Factor completely $a^2x - b^2x - a^2y + b^2y$.

$$a^2x - b^2x - a^2y + b^2y = x(a^2 - b^2) - y(a^2 - b^2)$$
$$= (a^2 - b^2)(x - y)$$
$$= (a + b)(a - b)(x - y)$$

Exercise 1.4.1

Factor completely:

1. $5a^2 - 10a$
2. $6a^2b + 9ab^2 - 6ab$
3. $9x^2 - 30x + 25$
4. $25x^2 - 49$
5. $5x^2 - 45$
6. $x^2 + 5x + 6$
7. $x^2 + 2x - 15$
8. $2x^2 + 5x + 2$
9. $2x^2 - x - 3$
10. $3x^2 + 13x - 10$
11. $6x^2 - 5x - 4$
12. $4x^2 + 14x + 6$
13. $6x^3 + 8x^2 - 8x$
14. $3x^2 + 9x + 15$
15. $5x^2 - 2x + 3$
16. $cx + dx + cy + dy$

17. $ax + 20 - 5a - 4x$
18. $a^2x - 8 + 2a^2 - 4x$
19. $2a^2x - 3 - 2x + 3a^2$
20. $27a^3 + 1$
21. $x^3 - y^6$
22. $a^9 - 1$
23. $2(x + 5)^2 + 3(x + 5) - 20$
24. $3(x - 2)^2 - 4(x - 2) - 7$
25. $x^2 + 2x - y^2 - 4y - 3$
26. $x^2 + 10x - 9y^2 + 6y + 24$
27. $9a^2 - (3x + 4y)^2$
28. $x^2 - 4x - y^2 - 6yz + 4 - 9z^2$
29. $18 - 2a^2 - 27b + 3a^2b$
30. $x^4 + 2acx^2 - w^4 - 2acw^2$

1.5 ALGEBRAIC FRACTIONS

Goals

Upon completion of this section you should be able to:

1. Add, subtract, multiply, and divide algebraic fractions.
2. Simplify complex algebraic fractions.

✳ ✳

In algebra, as in arithmetic, the operations on fractions are the basic operations of addition, subtraction, multiplication, and division, as well as simplifying. In this section these operations will be reviewed.

The *fundamental principle of fractions*

$$\frac{a}{b} = \frac{ax}{bx} \text{ for all } x \neq 0$$

is the basis of all operations on fractions.

Simplifying, multiplying, and dividing (i.e., multiplying by the inverse) are all a direct application of the fundamental principle of fractions. The simple rule, "factor completely all numerators and denominators, then cancel all like factors that are in both numerator and denominator" is sufficient for all such problems.

CAUTION: Only factors can be canceled. Never make the common error of canceling terms.

EXAMPLE

Simplify $\dfrac{ax + ay + 3x + 3y}{a^2 + 5a + 6}$.

$$\frac{ax + ay + 3x + 3y}{a^2 + 5a + 6} = \frac{\cancel{(a + 3)}(x + y)}{\cancel{(a + 3)}(a + 2)}$$

$$= \frac{x+y}{a+2}$$

EXAMPLE

Divide: $\dfrac{4x^2 - y^2}{x + 2y} \div 4x^2 - 2xy$.

First we change to a multiplication problem by using the inverse.

$$\frac{4x^2 - y^2}{x + 2y} \cdot \frac{1}{4x^2 - 2xy}$$

Then apply the rule.

$$\frac{\cancel{(2x-y)}(2x+y)}{x+2y} \cdot \frac{1}{2x\cancel{(2x-y)}} = \frac{2x+y}{2x(x+2y)}$$

In addition or subtraction of fractions, the fundamental principle is applied to change all fractions to those having like denominators; then the numerators are combined.

EXAMPLE

Combine $\dfrac{2x}{x+3} - \dfrac{5x+1}{x^2+x-6} + \dfrac{x}{x-2}$.

The lowest common denominator is $(x+3)(x-2)$.

$$\frac{2x}{x+3} - \frac{5x+1}{x^2+x-6} + \frac{x}{x-2} = \frac{2x(x-2) - (5x+1) + x(x+3)}{(x+3)(x-2)}$$

$$= \frac{3x^2 - 6x - 1}{(x+3)(x-2)}$$

The final answer for any problem should be left in simplified form. Sometimes after adding or subtracting, the fractional answer can be reduced.

Complex fractions are simplified by a direct application of the fundamental principle.

EXAMPLE

Simplify $\dfrac{\dfrac{-13}{x^2 - 2x - 35} - \dfrac{1}{x+5}}{\dfrac{1}{x+5} + 1}$.

The lowest common denominator of all fractions in the expression is $(x+5)(x-7)$. Thus to eliminate all individual fractions, we multiply

1.5 ALGEBRAIC FRACTIONS

both numerator and denominator by this common denominator.

$$\frac{(x+5)(x-7)\left[\dfrac{-13}{x^2-2x-35} - \dfrac{1}{x+5}\right]}{(x+5)(x-7)\left[\dfrac{1}{x+5} + 1\right]} = \frac{-13-(x-7)}{(x-7)+(x+5)(x-7)}$$

$$= \frac{-x-6}{x^2-x-42}$$

$$= \frac{-(x+6)}{(x+6)(x-7)}$$

$$= -\frac{1}{x-7} \text{ or } \frac{1}{7-x}$$

Exercise 1.5.1

Simplify the following:

1. $\dfrac{x^2 - 2x - 15}{x^2 + 3x - 40}$

2. $\dfrac{x^2 - 49}{x^2 + 14x + 49}$

3. $\dfrac{6x + 30}{10x^2 + 40x - 50}$

4. $\dfrac{3x + 6}{x} \cdot \dfrac{5x}{21x + 42}$

5. $\dfrac{x^2 - 1}{x + 1} \cdot \dfrac{x + 2}{x - 1}$

6. $\dfrac{x^2 + 3x + 2}{x^2 + 5x + 6} \cdot \dfrac{x^2 - 2x - 15}{x^2 - 3x - 10}$

7. $\dfrac{x + 1}{x^2 - 5x - 6} \div \dfrac{x^2 - 1}{x - 6}$

8. $\dfrac{2x - 6}{3x} \div \dfrac{x^2 - 2x - 3}{x^2 + x}$

9. $\dfrac{x^2 + 7x + 10}{x^2 + 4x - 5} \div \dfrac{x^2 + 7x + 12}{x^2 + 2x - 3}$

10. $\dfrac{x^2 - 16}{x^2 + 5x + 4} \div \dfrac{x^2 - 8x + 16}{x^2 - 1}$

11. $\dfrac{4}{x^2 + 3x} + \dfrac{x}{x + 3}$

12. $\dfrac{4}{x + 1} + \dfrac{1}{x^2 - 4} + \dfrac{5}{x + 2}$

13. $\dfrac{2x}{x^2 - 1} + \dfrac{3}{x^2 - x - 2}$

14. $\dfrac{5}{x - 1} - \dfrac{2x + 6}{x^2 + 2x - 3}$

15. $\dfrac{2x}{x^2 - 4} - \dfrac{x + 4}{x^2 + 4x - 12}$

16. $\dfrac{1}{x + 2} + \dfrac{1}{x + 1} - \dfrac{2}{x - 7}$

17. $\dfrac{1 - \dfrac{1}{a}}{a - 1}$

18. $\dfrac{\dfrac{1}{x} + \dfrac{1}{y}}{\dfrac{1}{x + y}}$

A BRIEF REVIEW OF THE BASIC PRINCIPLES OF ALGEBRA

19. $\dfrac{\dfrac{1}{x+3} - 1}{\dfrac{1}{x^2 - 9}}$

20. $\dfrac{\dfrac{6}{x^2 + 3x - 10} - \dfrac{1}{x-2}}{\dfrac{1}{x-2} + 1}$

CHAPTER REVIEW

Evaluate:

1. $|19 - 23|$
2. $\left|\dfrac{2}{3} - \dfrac{5}{7}\right|$

Justify each of the following by stating *one* of the field properties:

3. $5 + (7 + 3) = (7 + 3) + 5$
4. $(8 \cdot 3) \cdot 2 = 8 \cdot (3 \cdot 2)$
5. $5 \cdot \dfrac{1}{5} = 1$

6. Is the set of nonnegative integers closed under the operation of subtraction? Explain.

Simplify the following. Assume values of x and y which yield real numbers.

7. $(x^2 yz^2)^3 (-2xy^3 z)^2$
8. $(x^2 y^{-3})^{-2}$
9. $(3^0 x^{-2} y^3)^{-4}$
10. $\left(\dfrac{2^{-3}}{x^4}\right)^{-2} \left(\dfrac{-2x^3}{x^{-1}}\right)^3$
11. $\sqrt[6]{-64}$
12. $\sqrt[12]{x^2 y^8}$
13. $\sqrt[8]{16x^6 y^{10}}$
14. $\sqrt{50} + \sqrt[3]{2} - \sqrt{32} - \sqrt[3]{16}$
15. $\sqrt{2}\,(3\sqrt{6} - \sqrt{10}) - 2\sqrt{3}\,(\sqrt{15} + 2\sqrt{12})$
16. $(2\sqrt{3} + \sqrt{2})(2\sqrt{3} - 5\sqrt{2})$
17. $\dfrac{1}{\sqrt{12}}$
18. $\dfrac{3}{\sqrt{2} + \sqrt{3}}$
19. $\dfrac{-3a^2 b + 2ab}{b^2}$ if $a = -1, b = 3$
20. $21xyz + 15xy - 17xyz + 7xy^2$
21. $[4x - (x + 1)] - [3x + (x + 1)]$
22. $(x - 6)(x^2 + 6x + 36)$

Factor:

23. $15x^4 y - 20x^2 y^2 + 5x^2 y$
24. $64x^3 - 27y^3$
25. $x^2 + 16x + 15$
26. $x^2 - 20x + 96$

27. $x^2 + 23x - 140$

28. $4x^2 + 4x + 1$

29. $9x^2 - 12x + 4$

30. $15x^2 - 19x - 10$

31. $4x^3 + 8x^2 - 60x$

32. $xy - 8 - 2x + 4y$

33. $a^4x + 3 - x - 3a^4$

34. $4x^2 + 4x - y^2 - 4y - 3$

Simplify:

35. $\dfrac{3a^2 + 17a + 10}{2a^2 + 7a - 15}$

36. $\dfrac{3x^2 + 17x + 10}{x^2 + 10x + 25} \cdot \dfrac{x^2 - 25}{3x^2 + 11x + 6} \cdot \dfrac{5x^2 + 16x + 3}{5x^2 - 24x - 5}$

37. $\dfrac{25x^2 - 1}{10x^2 + 17x + 3} \div (1 - 5x)$

38. $\dfrac{x}{x^2 - 4x - 5} - \dfrac{2}{x + 1} + \dfrac{2x - 1}{x^2 - 7x + 10}$

39. $\dfrac{\dfrac{1}{a} + \dfrac{1}{a+b}}{\dfrac{2}{a+b} + 1}$

40. $\dfrac{\dfrac{1}{x^2 + 4x + 3} + \dfrac{1}{x - 2}}{\dfrac{1}{x + 3} + \dfrac{1}{x^2 - x - 2}}$

PRACTICE TEST

1. Evaluate $\left|\dfrac{3}{8} - \dfrac{4}{7}\right|$.

2. Which field property is illustrated by each of the following:

    a. $6 + (12 + 1) = 6 + (1 + 12)$

    b. $8 + 0 = 8$

    c. $3(2 + 9) = 3(2) + 3(9)$

3. Simplify the following. Assume values of x and y which yield real numbers.

    a. $(3^{-2})^3(3^0)^{-1}$

    b. $\sqrt[11]{-1}$

    c. $\sqrt[10]{16x^6y^8}$

    d. $\sqrt{8} - 2\sqrt{18} + \sqrt{50}$

    e. $4\sqrt{3}(\sqrt{6} - \sqrt{3} + \sqrt{18})$

    f. $\dfrac{2}{\sqrt{5} + 1}$

A BRIEF REVIEW OF THE BASIC PRINCIPLES OF ALGEBRA

5. Simplify:
   a. $16a^2b - 4a^2b + 5a^2b^2 - 7a^2b$
   b. $2(x + y) - [3x - (x + y)]$
   c. $(x - 5)(x^2 + 5x + 25)$

6. Factor:
   a. $15x^2 + 5x - 70$
   b. $3y - x - 3 + xy$
   c. $x^2 + 8xy + 16y^2 - 25$

7. Simplify the following. Assume the denominators do not equal zero.

   a. $\dfrac{x^2 - 2x - 63}{x^2 - 81} \div \dfrac{2x^2 - 5x - 12}{2x + 3}$

   b. $\dfrac{x + 2}{x - 5} - \dfrac{5x + 31}{x^2 - 2x - 15}$

   c. $\dfrac{\dfrac{1}{x + y} - \dfrac{1}{y}}{\dfrac{1}{x^2 - y^2}}$

PRACTICE TEST

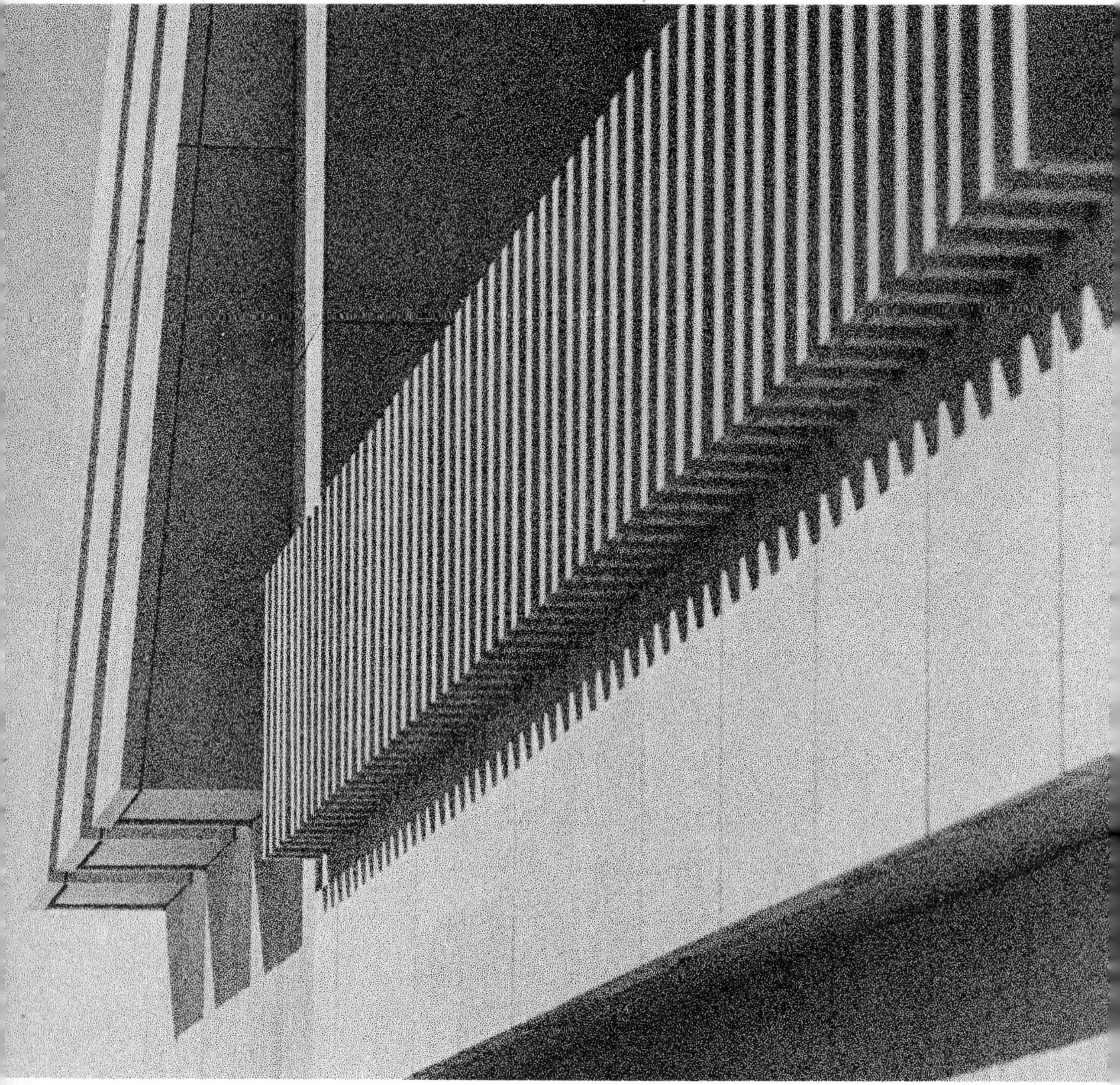

Equations and Inequalities

2

2

An *equation* is a statement indicating that two algebraic expressions are equal. An *inequality* is a statement indicating that one algebraic expression is less than another. The *solution set* of an equation or inequality is the set of numbers which, when substituted for the variable, will make the statement true. In this chapter, various types of equations and inequalities will be studied.

2.1 FIRST-DEGREE EQUATIONS— ONE VARIABLE

Goals

Upon completion of this section you should be able to:

1. Solve and check first-degree equations in one variable.
2. Solve and check first-degree equations involving algebraic fractions.

* *

The *degree* of an equation with one variable is the highest power of that variable found in the equation.

$$5x + 3(x - 2) = 7 \text{ is a first-degree equation}$$

$$3x^5 - 2x^3 + 4 = x \text{ is a fifth-degree equation}$$

$$x^2 + 4x - 2 = 0 \text{ is a second-degree equation}$$

Definition

Two equations are *equivalent* if they have the same solution set.

The process of solving equations involves changing a more complicated equation into an equivalent equation whose solution set is obvious. The operations that change an equation into an equivalent equation are

1. Adding the same quantity to both sides of the equation.
2. Subtracting the same quantity from both sides of the equation.

3. Multiplying or dividing both sides of the equation by the same non-zero quantity.

The following steps will, if followed in order, give the solution set for any first-degree equation:

1. Remove all parentheses.
2. Multiply both sides of the equation by a common denominator for all fractions in the equation.
3. Simplify by combining like terms on each side of the equation.
4. Add or subtract the quantities necessary to get the variable on one side and the other quantities on the other.
5. Divide both sides of the equation by the coefficient of the variable.
6. Substitute the solution into the original equation to verify that it is correct. If such a substitution makes the original statement true, then the solution is correct; otherwise it is incorrect.

EXAMPLE

Solve for x.

$$3(x + 1) + 7 = 2x - (x + 4)$$
$$3x + 3 + 7 = 2x - x - 4$$
$$3x + 10 = x - 4$$
$$2x = -14$$
$$x = -7$$

Check $3(-7 + 1) + 7 = -11$ and $2(-7) - (-7 + 4) = -11$. Since $x = -7$ makes the original statement true, the solution is correct.

EXAMPLE

Solve for y.

$$\frac{2}{3}(y - 3) + \frac{1}{2}y = 16 - \left(\frac{1}{4}y + 1\right)$$
$$\frac{2}{3}y - 2 + \frac{1}{2}y = 16 - \frac{1}{4}y - 1$$

Multiplying both sides by 12 gives

$$8y - 24 + 6y = 192 - 3y - 12$$
$$14y - 24 = 180 - 3y$$

2.1 FIRST-DEGREE EQUATIONS—ONE VARIABLE

$$17y = 204$$
$$y = 12$$

Substituting 12 for y in the original equation shows the answer to be correct.

Exercise 2.1.1

Solve the following:

1. $3(x - 4) = 5 + 2(x + 1)$
2. $3x + 2(x - 5) = 7 - (x + 3)$
3. $7x + 5 - 2(x - 1) = 21$
4. $\dfrac{x}{2} = \dfrac{1}{5} - x$
5. $x - \dfrac{1}{2} = \dfrac{x}{3} + 7$
6. $2x - 3(x - 2) = \dfrac{1}{2}(x + 1)$
7. $\dfrac{2}{3} + 1 = x - \dfrac{5}{2}$
8. $x - \dfrac{1}{5}x + 1 = \dfrac{1}{3}(x - 5)$
9. $\dfrac{2}{3}x - \dfrac{1}{2}x = x + \dfrac{1}{6}$
10. $\dfrac{x}{5} - \dfrac{2}{3}x + \dfrac{1}{2} = \dfrac{1}{3}(x - 4)$

Sometimes an equation that has a variable in the denominator of a fraction will reduce to a first-degree equation when multiplied by a common denominator. Such equations can then be solved by the step-by-step method discussed at the beginning of this section. However, one difficulty can arise. Since the common denominator contains the variable, we must be sure we are multiplying by a nonzero quantity.

EXAMPLE

Solve for x: $\dfrac{3}{x + 1} + 2 = 8$.

The lowest common denominator is $x + 1$, and we can multiply by this quantity only if $x \neq -1$. Doing so gives

$$3 + 2(x + 1) = 8(x + 1)$$
$$3 + 2x + 2 = 8x + 8$$
$$5 + 2x = 8x + 8$$
$$-6x = 3$$
$$x = -\dfrac{1}{2}$$

Substituting this value for x in the original equation shows the solution to be correct.

EXAMPLE

Solve for x: $\dfrac{2}{x^2 - 1} - \dfrac{1}{x + 1} = \dfrac{1}{x - 1}$.

The lowest common denominator is $x^2 - 1$, and multiplying by this quantity gives an equivalent equation only if $x \neq \pm 1$ (i.e., $x^2 - 1 \neq 0$). Multiplying gives

$$2 - (x - 1) = (x + 1)$$
$$2 - x + 1 = x + 1$$
$$-x + 3 = x + 1$$
$$-2x = -2$$
$$x = 1$$

Since $x \neq \pm 1$ and $x = 1$ are contradictory statements, the original equation has no solution.

Probably a simpler way to arrive at the type of conclusion shown in the last example is to solve the equation, then try to check. If the proposed solution makes any denominator equal to zero, there is no solution since a fraction with a zero denominator is meaningless.

Exercise 2.1.2

Solve the following:

1. $\dfrac{5}{x} = \dfrac{1}{2}$

2. $\dfrac{1}{x} + \dfrac{1}{2x} = 3$

3. $\dfrac{2}{x} + \dfrac{1}{3} = \dfrac{1}{2x} - 1$

4. $\dfrac{x + 1}{x} - \dfrac{x - 2}{2x} = 1$

5. $\dfrac{1}{x} = \dfrac{3}{2x}$

6. $\dfrac{4}{x} + \dfrac{1}{2x} = \dfrac{9}{4}$

7. $\dfrac{3}{x - 1} = 1$

8. $\dfrac{x}{x - 2} = \dfrac{4}{3}$

9. $\dfrac{2}{x} = \dfrac{1}{x + 1}$

10. $\dfrac{x}{x + 1} = \dfrac{x - 1}{x - 2}$

2.1 FIRST-DEGREE EQUATIONS—ONE VARIABLE

11. $\dfrac{1}{x+1} = \dfrac{2}{1-x^2}$ 12. $\dfrac{x+1}{x^2+2x-3} + \dfrac{1}{x-1} = \dfrac{1}{x+3}$

13. $\dfrac{x+2}{x^2+7x} = \dfrac{1}{x+3}$ 14. $\dfrac{1}{x^2-3x} = \dfrac{2}{x^2-9}$

15. $\dfrac{x}{x+1} + \dfrac{1}{2x+1} = 1$ 16. $\dfrac{3}{x+5} = 1 - \dfrac{x-4}{2x+10}$

17. $\dfrac{x}{x+4} - \dfrac{x}{x-4} = \dfrac{x+18}{x^2-16}$ 18. $\dfrac{2}{x-1} + \dfrac{x}{x+1} = 1 - \dfrac{1}{x^2-1}$

19. $\dfrac{2}{3x+6} = \dfrac{1}{6} - \dfrac{1}{2x+4}$ 20. $\dfrac{x+1}{x+5} - \dfrac{2x+1}{x-2} = \dfrac{5-x^2}{x^2+3x-10}$

2.2 EQUATIONS OF HIGHER DEGREE

Goals

Upon completion of this section you should be able to:

1. Arrange a quadratic equation in standard form.
2. Use the discriminant to determine the nature of the roots of a quadratic equation.
3. Solve quadratic equations by factoring or using the quadratic formula.

* *

The degree of a polynomial equation in x is the highest power of x appearing in the equation. Equations of degree higher than 1 are solved by various methods, and the real roots are estimated by various methods when the equation cannot be solved. At this point, solving by factoring will be discussed. In Chapter 10, other methods of finding or estimating roots will be discussed.

Solving equations by factoring is based on a theorem from algebra which says that "if a product is zero, then one of the factors is zero."

Theorem

If $ab = 0$, then $a = 0$ or $b = 0$.

This theorem can easily be proved. We start with the statement $ab = 0$ (called the hypothesis of the theorem). Now either $a = 0$ or $a \neq 0$. If $a = 0$, then the statement $a = 0$ or $b = 0$ is obviously true. If however $a \neq 0$, then we can divide each side of $ab = 0$ by a and obtain $b = 0$. Hence we have established the theorem.

Since our ability to factor is limited mostly to second-degree polynomials, second-degree (quadratic) equations will be the center of our discussion. Later in this chapter we will consider equations which are quadratic in form but which are not quadratic equations themselves.

Definition The standard form of a quadratic equation is

$$ax^2 + bx + c = 0$$

where $a \neq 0$ and a, b, c are real numbers.

Factorable equations can be solved by the following steps:

1. Arrange the equation in standard form.
2. Factor the polynomial completely.
3. Set each factor equal to zero and solve the resulting equations for the variable.
4. Substitute each value of the variable into the original equation to determine if it is a solution.

Step 4 is especially necessary when the original equation is not in the form of a polynomial.

EXAMPLE   Solve for x.

$$x^2 - 5x = 0$$
$$x(x - 5) = 0$$
$$x = 0 \quad \text{or} \quad x - 5 = 0$$
$$x = 0 \quad \text{or} \quad x = 5$$

Both solutions check, so the solution set is $\{0, 5\}$.

EXAMPLE   Solve for x.

$$5x^2 + 13x = -6$$
$$5x^2 + 13x + 6 = 0$$
$$(5x + 3)(x + 2) = 0$$
$$5x + 3 = 0 \quad \text{or} \quad x + 2 = 0$$
$$x = -\frac{3}{5} \quad \text{or} \quad x = -2$$

Both solutions check, so the solution set is $\left\{-\frac{3}{5}, -2\right\}$.

2.2 EQUATIONS OF HIGHER DEGREE

EXAMPLE Solve for x.

$$3x^2 + xy - 4y^2 = 0$$
$$(3x + 4y)(x - y) = 0$$
$$3x + 4y = 0 \quad \text{or} \quad x - y = 0$$
$$x = -\frac{4y}{3} \quad \text{or} \quad x = y$$

Both solutions check, so the solution set is $\left\{-\frac{4y}{3}, y\right\}$.

EXAMPLE Solve for x.

$$6x^3 + 17x^2 + 7x = 0$$
$$x(6x^2 + 17x + 7) = 0$$
$$x(2x + 1)(3x + 7) = 0$$
$$x = 0 \quad \text{or} \quad 2x + 1 = 0 \quad \text{or} \quad 3x + 7 = 0$$
$$x = 0 \quad \text{or} \quad x = -\frac{1}{2} \quad \text{or} \quad x = -\frac{7}{3}$$

All solutions check, so the solution set is $\left\{0, -\frac{1}{2}, -\frac{7}{3}\right\}$.

EXAMPLE Solve for x: $\dfrac{5}{2x+1} + \dfrac{1}{x+2} = -2.$

Multiplying by $(2x + 1)(x + 2)$ gives
$$5(x + 2) + (2x + 1) = -2(2x + 1)(x + 2)$$
$$5x + 10 + 2x + 1 = -4x^2 - 10x - 4$$
$$4x^2 + 17x + 15 = 0$$
$$(x + 3)(4x + 5) = 0$$
$$x + 3 = 0 \quad \text{or} \quad 4x + 5 = 0$$
$$x = -3 \quad \text{or} \quad x = -\frac{5}{4}$$

The first step, multiplying by $(2x + 1)(x + 2)$, i.e., the lowest common

EQUATIONS AND INEQUALITIES

denominator, requires that $x \neq -\frac{1}{2}$ and $x \neq -2$, since either of these values would mean $(2x + 1)(x + 2) = 0$. In solving an equation such as this example, if either solution causes a denominator to be zero, that solution is rejected. A solution might check in the standard form but not in the original. It is therefore important that the original equation be used to check all solutions.

The solution set is $\left\{-3, -\frac{5}{4}\right\}$.

EXAMPLE

Solve for x: $\dfrac{x^2 + 2}{x - 2} = 4 + \dfrac{3x}{x - 2}$.

Multiply by $x - 2$ and arrange in standard form.
$$x^2 - 7x + 10 = 0$$
$$(x - 5)(x - 2) = 0$$
$$x - 5 = 0 \quad \text{or} \quad x - 2 = 0$$
$$x = 5 \quad \text{or} \quad x = 2$$

In checking, $x = 2$ makes the denominator equal to zero, but $x = 5$ checks, so the solution set is $\{5\}$.

Exercise 2.2.1

Solve for x by factoring.

1. $2x^2 + 6x = 0$
2. $2x^2 - 3x = 0$
3. $x^2 + 2x = 3$
4. $2x^2 + 5x + 3 = 0$
5. $x^2 + 9 = 6x$
6. $x = 2 + \dfrac{35}{x}$
7. $5x + \dfrac{4}{3x} = \dfrac{23}{3}$
8. $6x^3 - 13x^2 = 5x$
9. $3x^2 + 5xy = 2y^2$
10. $\dfrac{1}{x + 2} + 2 = \dfrac{3}{x^2 + 2x}$
11. $\dfrac{2x^2 + 3}{x + 1} = 6 - \dfrac{5x}{x + 1}$
12. $\dfrac{4}{x + 2} + \dfrac{1}{2x - 5} = 1$

Quadratic equations that are not factorable can be solved by using the *quadratic formula*. This formula is developed from the standard form of the quadratic equation by the method of *completing the square.*

A perfect square trinomial $(x + k)^2$ is $x^2 + 2kx + k^2$ and from this form a specific relationship is evident. When the coefficient of x^2 is 1, the constant term is the square of half the coefficient of the x term. This fact is used to complete the square on the general quadratic in the following manner.

$$ax^2 + bx + c = 0$$

First we divide by a so that the coefficient of x^2 will be 1.

$$x^2 + \frac{b}{a}x + \frac{c}{a} = 0$$

Now, since the coefficient of the x term must be twice the square root of the constant term, we will supply a constant term to fulfill this requirement. We take one-half the coefficient of x, giving $\frac{b}{2a}$, then add the square of this as a constant term to each side of the equation.

$$x^2 + \frac{b}{a}x + \frac{b^2}{4a^2} = \frac{b^2}{4a^2} - \frac{c}{a}$$

Factoring gives

$$\left(x + \frac{b}{2a}\right)^2 = \frac{b^2}{4a^2} - \frac{c}{a}$$

$$\left(x + \frac{b}{2a}\right)^2 = \frac{b^2 - 4ac}{4a^2}$$

We now use the fact that if $m^2 = n$, then $m = \pm \sqrt{n}$ and get

$$x + \frac{b}{2a} = \pm\sqrt{\frac{b^2 - 4ac}{4a^2}}$$

or

$$x = \frac{-b \pm \sqrt{b^2 - 4ac}}{2a}$$

which is the quadratic formula. This formula is developed from the general quadratic, and therefore, any equation which can be put into standard form can be solved by this formula. You should memorize the quadratic formula.

EXAMPLE

Solve $5x^2 - 7x = 8$ by the quadratic formula. Arrange the equation in standard form.

$$5x^2 - 7x - 8 = 0$$

Thus $\qquad a = 5 \qquad b = -7 \qquad c = -8$

Substituting these values in the quadratic formula gives

$$x = \frac{7 \pm \sqrt{49 + 160}}{10}$$

or

$$x = \frac{7 \pm \sqrt{209}}{10}$$

and the solution set is $\left\{ \dfrac{7 + \sqrt{209}}{10}, \dfrac{7 - \sqrt{209}}{10} \right\}$.

It should be obvious from the formula that $b^2 - 4ac$ (the portion under the radical) will determine the nature of the roots of a quadratic. $b^2 - 4ac$ is called the *discriminant* because it gives the nature of the roots.

Examine the following table and determine why each entry is valid.

Roots of a Quadratic

value of the discriminant	nature of the roots (for a, b, c, rational numbers)
$b^2 - 4ac < 0$	No real roots.
$b^2 - 4ac = 0$	Roots are real and equal. (Two identical roots.) The trinomial is a perfect square.
$b^2 - 4ac > 0$ but not a perfect square	Roots are irrational and unequal. (Two different roots which contain a radical.)
$b^2 - 4ac > 0$ and a perfect square	Roots are unequal and rational. (Two different roots containing no radical.) (The trinomial is factorable.)

Exercise 2.2.2

Solve the following by the quadratic formula. Leave all solutions in simplest form.

1. $x^2 + 2x - 15 = 0$
2. $5x^2 = 7x + 6$
3. $2x^2 + 3x = 0$
4. $x^2 + 3x + 1 = 0$

2.2 EQUATIONS OF HIGHER DEGREE

5. $5x^2 + 7x + 1 = 0$ 6. $3x^2 - 5 = 0$

7. $x^2 + 2x = 7$ 8. $9x^2 + 12x + 4 = 0$

9. $3x^2 + 8x = 5$ 10. $3x^2 + 4x - 5 = 0$

Compute $b^2 - 4ac$ for each of the following and give the nature of the roots.

11. $x^2 - 7x + 12 = 0$ 12. $x^2 - 4x + 4 = 0$

13. $x^2 - 3x + 1 = 0$ 14. $2x^2 - 7x + 3 = 0$

15. $x^2 - 3x + 5 = 0$ 16. $5x^2 + 6x + 1 = 0$

17. $6x^2 + 11x - 21 = 0$ 18. $4x^2 + 4x + 1 = 0$

19. $2x^2 + 5x - 3 = 0$ 20. $x^2 + 5x + 7 = 0$

2.3 COMPLEX NUMBERS

Goals

Upon completion of this section you should be able to:

1. Add, subtract, multiply, and divide complex numbers.
2. Solve and check quadratic equations having complex numbers as solutions.

* *

One entry in the "Nature of Roots" table in the preceding section shows "no real roots." To the observant student this should imply that roots might exist in some other set of numbers, and such is the case.

Suppose, for example, that we want to find a solution to the equation $x^2 + 1 = 0$. This means that we must find a value for x such that, when we square it, the result will be -1. Can you see why no real number would satisfy this condition? If we want a solution, therefore, we must look to another set of numbers.

To introduce this new set of numbers, we define the imaginary unit i as the square root of -1.

Definition $i = \sqrt{-1}$ or $i^2 = -1$

Accepting this definition of i makes it possible to find values for the square roots of negative numbers, or at least to indicate such values.

EXAMPLE Find the value of $\sqrt{-4}$.

$$\sqrt{-4} = \sqrt{(-1)(4)}$$
$$= \sqrt{-1}\sqrt{4}$$
$$= i\sqrt{4}$$
$$= 2i$$

To check $\sqrt{-4} = 2i$ by the definition of square root, we evaluate $(2i)^2$.

$$(2i)^2 = (2i)(2i)$$
$$= (2)(2)(i)(i)$$
$$= 4i^2$$

Since $i^2 = -1$, then $4i^2 = -4$.

EXAMPLE Find the value of $\sqrt{-4}\sqrt{-4}$. We must proceed as follows:

$$\sqrt{-4}\sqrt{-4} = (2i)(2i) = 4i^2 = -4$$

which is the *correct* value.

> **CAUTION:** Many students fall into the trap of writing $\sqrt{-4}\sqrt{-4} = \sqrt{(-4)(-4)} = \sqrt{16} = 4$. This is *not* accurate since $\sqrt{a}\sqrt{b} = \sqrt{ab}$ is true for *real numbers* only.

EXAMPLE Find the value of $\sqrt{-10}$.

$$\sqrt{-10} = \sqrt{(-1)(10)}$$
$$= \sqrt{-1}\sqrt{10}$$
$$= i\sqrt{10}$$

(The answer is left in this form since $\sqrt{10}$ is not rational.)

A number such as $2i$ or $i\sqrt{10}$ is called an *imaginary number*.

Definition The indicated sum of a real number and an imaginary number $(a + bi)$ where a and b are real is called a *complex number*.

Definition   Two complex numbers $(a + bi)$ and $(c + di)$ are *equal* if and only if $a = c$ and $b = d$.

It should first be noted that the set of real numbers is a subset of the set of complex numbers. A real number x can be expressed as $x + 0i$ and so, by definition, is complex. Also the set of imaginary numbers is a subset of the set of complex numbers since an imaginary number bi can be expressed as $0 + bi$.

Complex numbers form a field. They can be added, subtracted, multiplied, divided, raised to powers, etc., as you would expect of any set of numbers that forms a field. Rules for these operations follow.

Rule   To add or subtract complex numbers, combine the real parts and combine the imaginary parts. (Note that this follows a previous rule that states that only like terms can be combined.)

EXAMPLE   Add: $(7 + 6i) + (3 + 2i)$.

$$(7 + 6i) + (3 + 2i) = (7 + 3) + (6i + 2i)$$
$$= 10 + 8i$$

EXAMPLE   Subtract: $(3 + 4i) - (2 - 6i)$.

$$(3 + 4i) - (2 - 6i) = 3 + 4i - 2 + 6i$$
$$= 1 + 10i$$

Rule   To multiply complex numbers, consider them as binomials and use the distributive property.

EXAMPLE   Multiply: $(3 + 2i)(4 + 6i)$.

$$(3 + 2i)(4 + 6i) = 12 + 26i + 12i^2$$
$$= 12 + 26i + 12(-1)$$
$$= 26i$$

EXAMPLE   Multiply: $5(3 + 4i)$.

$$5(3 + 4i) = 15 + 20i$$

This is a direct application of the distributive property.

The next operation we will discuss is division. Division is always defined as multiplying by the inverse. Hence, we need to find the inverse of a complex number.

The field property on multiplicative inverses states basically that the product of a number and its inverse is 1. We need, then, to find x in the equation

$$(a + bi)(x) = 1$$

Immediately we obtain $\qquad x = \dfrac{1}{a + bi}$

This expression for the inverse must be simplified since $i = \sqrt{-1}$ is a radical in the denominator of a fraction.

$$\frac{1}{a + bi} = \frac{1}{a + bi} \cdot \frac{a - bi}{a - bi}$$

$$= \frac{a - bi}{a^2 + b^2}$$

Therefore, the *inverse* of $a + bi$ is $\dfrac{(a - bi)}{(a^2 + b^2)}$. It can similarly be shown that the inverse of $a - bi$ is $\dfrac{(a + bi)}{(a^2 + b^2)}$.

Two complex numbers of the form $(a + bi)$ and $(a - bi)$ are called *conjugates*. For example, $(7 + 2i)$ and $(7 - 2i)$ are conjugates of each other.

Rule   **To divide by a complex number, multiply by its inverse.**

EXAMPLE   Divide: $(3 + 4i) \div (2 + i)$.

$$(3 + 4i) \div (2 + i) = (3 + 4i) \cdot \frac{(2 - i)}{(2^2 + 1^2)}$$

2.3 COMPLEX NUMBERS

$$= \frac{6 + 5i - 4i^2}{5}$$

$$= 2 + i$$

Exercise 2.3.1

Perform the following operations:

1. $(2 + 3i) + (5 + 4i)$
2. $(6 + 5i) + (1 - 6i)$
3. $(5 - 2i) - (8 - i)$
4. $(11 + 7i) - (3 + 4i)$
5. $(11 + i) - 3(2 - 5i)$
6. $(2 + 3i)(1 + 4i)$
7. $(x + iy)(x - iy)$
8. $(5 + 4i)^2$
9. $(6 - 5i)^2$
10. $(2 + 3i) \div (1 - 2i)$
11. $5 \div (4 + 3i)$
12. $2i \div (6 + i)$

The most common use of complex numbers at this level of algebra is in expressing solutions to equations, especially quadratics that have no real roots.

EXAMPLE

Solve $x^2 - 2x + 5 = 0$. Since the expression on the left will not factor, we will use the quadratic formula with $a = 1$, $b = -2$, and $c = 5$.

$$x = \frac{2 \pm \sqrt{(-2)^2 - 4(1)(5)}}{2}$$

$$= \frac{2 \pm \sqrt{-16}}{2}$$

$$= 1 \pm 2i$$

Check $x = 1 + 2i$.

$$(1 + 2i)^2 - 2(1 + 2i) + 5 = 1 + 4i + 4i^2 - 2 - 4i + 5$$
$$= 4 + 4i^2$$
$$= 0$$

Check $x = 1 - 2i$.

$$(1 - 2i)^2 - 2(1 - 2i) + 5 = 1 - 4i + 4i^2 - 2 + 4i + 5$$
$$= 4 + 4i^2$$

$$= 0$$

Therefore the solution set is $\{1 + 2i, 1 - 2i\}$.

EXAMPLE

Solve for x.

$$x^3 = 8$$
$$x^3 - 8 = 0$$
$$(x - 2)(x^2 + 2x + 4) = 0$$
$$x - 2 = 0 \quad \text{or} \quad x^2 + 2x + 4 = 0$$

$x - 2 = 0$ gives one solution, $x = 2$. $x^2 + 2x + 4 = 0$ will not factor. Using the quadratic formula with $a = 1$, $b = 2$, and $c = 4$, we obtain

$$x = \frac{-2 \pm \sqrt{4 - 16}}{2}$$
$$= -1 \pm i\sqrt{3}$$

Checking, we find that each of the three solutions is valid. Therefore the solution set is $\{2, -1 + i\sqrt{3}, -1 - i\sqrt{3}\}$.

The student should notice that in this example, we have actually found three cube roots of 8. One is real and two are complex. In Chapter 10, we will establish the fact that every real number has two square roots, three cube roots, four fourth roots, etc.

Exercise 2.3.2

Solve and check.

1. $x^2 - 4x + 8 = 0$
2. $2x^2 - 2x + 1 = 0$
3. $8x^2 - 4x + 1 = 0$
4. $x^2 + x + 1 = 0$
5. $x^2 - x + 1 = 0$
6. $x^2 - 2x + 4 = 0$
7. $x^3 - 1 = 0$
8. $x^3 + 8 = 0$
9. $x^3 - 27 = 0$
10. $x^3 + 27 = 0$
11. $x^4 = 16$
12. Is $(-2 + 2i\sqrt{3})$ a cube root of 64?
13. We know that $5 < 7$. Do you think we could say $(2 + 3i) < (3 + 4i)$? Explain.

2.4 EQUATIONS INVOLVING QUADRATIC FORM, RADICALS, AND ABSOLUTE VALUE

Goals

Upon completion of this section you should be able to:

1. Recognize when an equation is quadratic in form.
2. Solve equations quadratic in form.
3. Solve equations involving radicals.
4. Solve equations involving absolute value.

* *

In the preceding sections we have discussed first-degree equations, equations that are factorable, and quadratic equations. There are certain other types of equations that can be solved by various methods which we will discuss in this section.

Definition

An equation is *quadratic in form* if a substitution can be found for the variable that will result in an equation of the form $ax^2 + bx + c = 0$.

EXAMPLE

The equation $x^{\frac{2}{3}} + 5x^{\frac{1}{3}} = 6$ is quadratic in form, since the substitution

$$y = x^{\frac{1}{3}}$$

results in the equation $\quad y^2 + 5y - 6 = 0$

Rule

To solve an equation that is quadratic in form, make the suitable substitution, solve the resulting quadratic equation, and then use these solutions together with the substitution to find possible roots and check them in the original equation.

EXAMPLE

Solve for x: $x^{\frac{2}{3}} + 5x^{\frac{1}{3}} = 6$. If we let

$$y = x^{\frac{1}{3}}$$

then the equation becomes $\quad y^2 + 5y = 6$
which when solved yields

$$y = -6 \quad \text{or} \quad y = 1$$

Using the original substitution, we have

$$x^{\frac{1}{3}} = -6 \quad \text{or} \quad x^{\frac{1}{3}} = 1$$

Therefore
$$x = (-6)^3 \quad \text{or} \quad x = (1)^3$$
$$x = -216 \quad \text{or} \quad x = 1$$

Checking these solutions in the original equation verifies that they are roots. The solution set is $\{-216, 1\}$.

EXAMPLE

Solve for x: $3x^4 - 2x^2 - 5 = 0$. Let $y = x^2$.

Then
$$3y^2 - 2y - 5 = 0$$

giving
$$y = \frac{5}{3} \quad \text{or} \quad y = -1$$

Thus
$$x^2 = \frac{5}{3} \quad \text{or} \quad x^2 = -1$$

and
$$x = \pm\frac{\sqrt{15}}{3} \quad \text{or} \quad x = \pm i$$

The solution set is $\left\{\pm\frac{\sqrt{15}}{3}, \pm i\right\}$.

Another type of equation that requires special treatment is the equation containing radicals. To solve such an equation, we must eliminate the radical by raising both sides of the equation to a common power. This process must be repeated until we have an equation free of radicals or in a form that we can solve by a suitable substitution. The solutions *must* be checked in the original equation.

EXAMPLE

Solve $\sqrt{x + 2} = x - 4$. Squaring both sides we have

$$x + 2 = x^2 - 8x + 16$$
$$x^2 - 9x + 14 = 0$$

This solves giving $\quad x = 7 \quad \text{or} \quad x = 2$

Checking these solutions in the original equation, we find that $x = 7$ checks but $x = 2$ does not check. Therefore, the solution set is $\{7\}$.

2.4 EQUATIONS/QUADRATIC FORM, RADICALS, AND ABSOLUTE VALUE

In the preceding example, $x = 2$ is called an *extraneous root*. Extraneous roots are roots that appear in solution methods but do not check in the original equation. Can you determine the step that "picked up" the extra root? Can you explain why this happened? Solutions must always be checked in the original equation when working with radicals! Sometimes all roots will be extraneous, in which case we would have no solution.

EXAMPLE

Solve $\sqrt{x + 2} = \sqrt{7 - x} - 3$. Since we cannot isolate the radical, we square as is, giving

$$x + 2 = (7 - x) - 6\sqrt{7 - x} + 9$$
$$2x - 14 = -6\sqrt{7 - x}$$

Squaring again gives

$$4x^2 - 56x + 196 = 36(7 - x)$$

which simplifies to

$$x^2 - 5x - 14 = 0$$

This solves to give

$$x = 7 \quad \text{or} \quad x = -2$$

Checking in the original equation, we find that the solution set is $\{-2\}$.

Another special type of equation is one containing absolute value. To solve equations involving absolute value, we must remember the definition

$$|x| = \begin{cases} x & \text{if } x \geq 0 \\ -x & \text{if } x < 0 \end{cases} \quad \text{for all real } x$$

EXAMPLE

Solve for x: $|x + 5| = 7$. From the definition we see that this equation can be interpreted as

$$x + 5 = 7 \quad \text{or} \quad x + 5 = -7$$

since in either case, $|x + 5| = 7$. Solving each of the resulting equations, we find the solution set to be $\{-12, 2\}$.

EXAMPLE

Solve for x: $|2x^2 - x - 6| = 4$.

$$2x^2 - x - 6 = 4 \quad \text{or} \quad 2x^2 - x - 6 = -4$$

Solving each of these quadratic equations, we find the solution set to be

$$\left\{-2, \frac{5}{2}, \frac{1+\sqrt{17}}{4}, \frac{1-\sqrt{17}}{4}\right\}$$

Exercise 2.4.1

Solve for x.

1. $x^4 - 5x^2 + 4 = 0$
2. $x^4 - 6x^2 + 8 = 0$
3. $x^4 - 7x^2 + 10 = 0$
4. $x^4 - 3x^2 - 4 = 0$
5. $x^4 + 8x^2 + 15 = 0$
6. $x^6 - 7x^3 - 8 = 0$
7. $x^{\frac{1}{2}} - 5x^{\frac{1}{4}} + 6 = 0$
8. $x + \sqrt{x} = 6$
9. $x - 2\sqrt{x} = 3$
10. $x + \sqrt{2x - 5} = 10$
11. $x - 2\sqrt{2x + 1} = -2$
12. $5 - \sqrt{5x - 1} = x$
13. $\sqrt{x + 2} + \sqrt{2x - 10} = 5$
14. $\sqrt{x + 1} + \sqrt{2x + 3} = 5$
15. $\sqrt{x + 4} + \sqrt{2x + 10} = 3$
16. $|x + 3| = 8$
17. $|5x - 4| = 11$
18. $|2x + 7| = 7$
19. $|x^2 + 3x - 8| = 2$
20. $|2x^2 + x - 3| = 3$
21. $x - 4x^{\frac{1}{2}} - 21 = 0$
22. $x^{\frac{2}{3}} + 2x^{\frac{1}{3}} - 3 = 0$
23. $2x^{\frac{1}{2}} + 7x^{\frac{1}{4}} + 3 = 0$
24. $\sqrt{x + 1} + \sqrt{x - 7} = \sqrt{2x}$
25. $\sqrt{2x + 5} - \sqrt{x + 2} = \sqrt{3x - 5}$
26. $|x^2 + 5| = 1$
27. $|3x + 4| = -2$
28. $|x^2 + 2x + 1| = 4$
29. $|3x - 4| = |x + 6|$
30. $|2x^2 + 2x - 1| = |x^2 - 4x - 6|$

2.5 FIRST-DEGREE INEQUALITIES

Goals

Upon completion of this section you should be able to:
1. Solve and graph first-degree inequalities.

* *

In Sec. 1.1 the number line and the inequality symbol were discussed. In this section we wish to solve first-degree inequalities and graph their solutions. A graph is a geometric representation of an algebraic statement and the number line is used to graph all one-dimensional solutions.

Some symbolism is necessary at this point, since there are several accepted methods of designating solutions on a one-dimensional graph. We will use interval notation that utilizes brackets [ ], parentheses ( ), and combinations of these.

To denote an interval on the number line when the endpoints are included, we will use the bracket. For example $[-5, 3]$ designates the interval from -5 to 3 including the numbers -5 and 3. To say x is in $[-5, 3]$, sometimes written $x \in [-5, 3]$, would mean the same as $-5 \leq x \leq 3$.

When the endpoints of the interval are not included, we will use parentheses. For example $(-5, 3)$ designates the interval from -5 to 3 excluding the numbers -5 and 3. To say x is in $(-5, 3)$ would mean the same as $-5 < x < 3$.

When the endpoints of an interval are included in the set, the interval is said to be a *closed interval*. When the endpoints are excluded, the interval is called an *open interval*. Of course we can have "half-open" or "half-closed" intervals. In this case one endpoint is included and the other is not.

The same notation (brackets and parentheses) can be used on the number line to graph a solution to an inequality. For instance, the statement $x \in [-5, 3]$ has the following graph:

and the statement $x \in (-5, 3)$ has the graph

The graph of $x \in (-5, 3]$ is

EQUATIONS AND INEQUALITIES

Exercise 2.5.1

Write the following in interval notation:

1. $3 \leqslant x \leqslant 8$
2. $-5 \leqslant x \leqslant 4$
3. $-1 < x < 6$
4. $0 \leqslant x < 9$
5. $-4 < x \leqslant -2$

Graph the following:

6. $x \in [-4, 8]$
7. $x \in (2, 7)$
8. $x \in (-2, 5]$
9. $x \in [-6, -3)$
10. How many real numbers are in the interval $(1, 2)$?

The rules for solving first-degree equations given in Sec. 2.1 will also solve first-degree inequalities. One difference must be noted which is extremely important. The following theorem must be taken into account as we solve first-degree inequalities.

Theorem   If $a < b$, then $-a > -b$.

This theorem implies that if we multiply or divide each side of an inequality by a negative quantity, the sense (direction) of the inequality will be reversed. The theorem is easily proved by adding $(-a) + (-b)$ to both sides of the inequality $a < b$.

Proof
$$a < b$$
$$(-a) + (-b) + a < b + (-a) + (-b)$$
$$-b < -a$$
$$-a > -b$$

EXAMPLE   Solve for x and graph the solution.

$$\frac{1}{2}(x + 3) + 2x < 3x - 1$$

$$\frac{1}{2}x + \frac{3}{2} + 2x < 3x - 1$$

$$x + 3 + 4x < 6x - 2$$

2.5   FIRST-DEGREE INEQUALITIES

$$-x < -5$$
$$x > 5 \quad \text{Why?}$$

The graph of the solution is

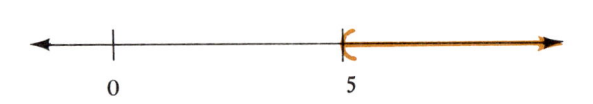

EXAMPLE

Solve for x and write the solution set in interval notation.

$$3(x + 1) + 2x < x - 5(x + 4)$$
$$3x + 3 + 2x < x - 5x - 20$$
$$9x < -23$$
$$x < -\frac{23}{9}$$

The solution in interval notation is $\left(-\infty, -\frac{23}{9}\right)$.

The symbol ∞ (read as "infinity") is *not* to be considered a real number. $(-\infty, +\infty)$ designates the set of all real numbers, $(-\infty, 0)$ the negative real numbers, and $(0, +\infty)$ the positive real numbers.

If we think of the absolute value of a number as its distance from zero, without regard to direction, then the statement $|x| < a$ would imply that x is less than a units from zero and $|x| > a$ would imply that x is more than a units from zero. Hence the following theorems.

Theorem $|x| < a \Leftrightarrow -a < x < a$. (The symbol $\Leftrightarrow$ is usually read as "if and only if.")

Theorem $|x| > a \Leftrightarrow x > a \quad$ or $\quad x < -a$

EXAMPLE

Solve for x: $|2x + 1| < 5$.

$$-5 < 2x + 1 < 5 \quad \text{Why?}$$
$$-6 < 2x < 4$$
$$-3 < x < 2$$

In interval notation the solution set is $(-3, 2)$, which graphs as

$$\xleftarrow{\qquad}\overset{\displaystyle (}{\underset{-3}{\quad}}\overset{}{\underset{0}{\quad}}\overset{\displaystyle)}{\underset{2}{\quad}}\xrightarrow{\qquad}$$

EXAMPLE

Solve for x: $|5x + 3| \geq 12$.

$$5x + 3 \geq 12 \quad \text{or} \quad 5x + 3 \leq -12 \quad \text{Why?}$$
$$5x \geq 9 \qquad\qquad\qquad 5x \leq -15$$
$$x \geq \frac{9}{5} \qquad\qquad\qquad x \leq -3$$

In interval notation the solution set is $(-\infty, -3] \cup \left[\frac{9}{5}, +\infty\right)$, since either interval makes the original inequality true.

The symbol $\cup$ is the set notation for union and corresponds to the logical "or." Therefore, the solution set could be read $(-\infty, -3]$ or $\left[\frac{9}{5}, +\infty\right)$.

Exercise 2.5.2

Solve the following and give solutions in interval form:

1. $4x + 3 < 2(x + 1)$
2. $6x - 5 \geq 10x + 3(x - 1)$
3. $\frac{1}{2}x + 3 < \frac{2}{3}x - 1$
4. $\frac{2}{5}(x + 1) \geq \frac{1}{4}x - \frac{1}{2}(x + 3)$
5. $x - 3 > 5\left(x - \frac{3}{5}\right)$
6. $\frac{1}{3}(3x + 5) \geq \frac{2}{3}(x + 1) - \frac{1}{2}(4x - 5)$
7. $|x| < 3$
8. $|4x + 3| < 11$
9. $|7x - 5| \geq 9$
10. $|3x + 2| \leq 0$

Solve the following and graph their solution sets:

11. $5x - 1 > 3(x + 1)$
12. $3(x + 3) + 2x > 3 + 7x$
13. $2(x + 3) \leq 7(x + 2) + 2$
14. $x > 4\left(\frac{1}{3}x + \frac{1}{2}\right)$
15. $\frac{2}{3}x + 1 \geq 2x - \left(\frac{x}{2} - 6\right)$
16. $5(x - 2) < 4(2x - 3) + 2$

17. $|x| > 2$
18. $|5x + 2| < 8$
19. $|3x - 4| \geq 5$
20. $|2x + 7| \leq 1$

2.6 OTHER INEQUALITIES IN ONE VARIABLE

Goals

Upon completion of this section you should be able to:

1. Solve and graph inequalities involving products and quotients of factors.

* *

In Sec. 2.2 we solved quadratic and other equations by factoring and by using the quadratic formula. Solving quadratic and other inequalities proceeds in the same manner, until we must interpret the solution set. For instance, if we have the inequality

$$x^2 + 5x + 6 < 0$$

and factor it as

$$(x + 3)(x + 2) < 0$$

we must recognize that the solution set will be those real numbers that will make the product of two factors negative. This means, of course, that one factor must be positive and the other negative. Several methods can be used to find the solution set but all are basically the same. We must find the real numbers such that

Case I $x + 3 > 0$ and $x + 2 < 0$

or

Case II $x + 3 < 0$ and $x + 2 > 0$

The method we will present here is a graphical method which will give us a visual interpretation of the cases and perhaps make the solution set more meaningful.

EXAMPLE

Solve for x: $x^2 + 5x + 6 < 0$.

Step 1 Factor obtaining

$$(x + 3)(x + 2) < 0$$

Step 2 Make a graph, on parallel number lines, of each factor being greater than zero.

EQUATIONS AND INEQUALITIES

$x + 3 > 0$

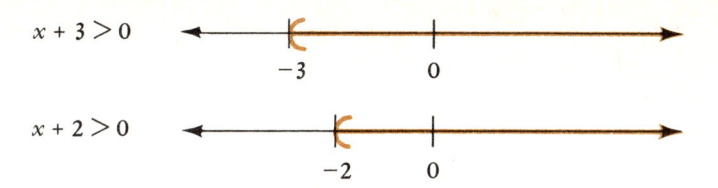

$x + 2 > 0$

Step 3 Project the intervals from the graphs of Step 2 onto a number line parallel to those graphs.

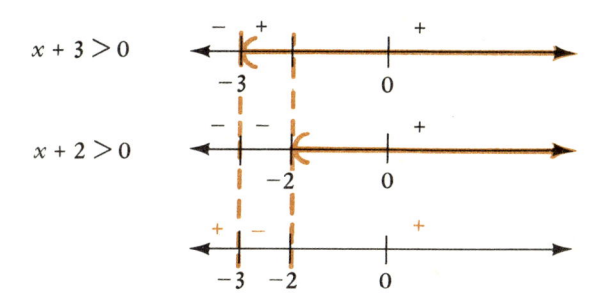

$x + 3 > 0$

$x + 2 > 0$

If we now recognize that in each case the graph shows where a factor is positive and where it is negative, we can project where the product will be positive and negative. We also know that the value of the endpoints themselves will make a factor, and hence the product, equal to zero. So the graph of the solution set is

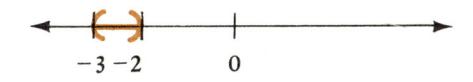

In interval notation the solution set is $(-3, -2)$.

EXAMPLE   Solve for x: $\dfrac{2x^2 + 17}{x + 4} > 5$.

In all inequalities, the first step is to make the right side equal to zero, since we then are looking for a set of numbers which will make a quotient or product either positive or negative.

$$\frac{2x^2 + 17}{x + 4} - 5 > 0$$

2.6 OTHER INEQUALITIES IN ONE VARIABLE

CAUTION: We cannot multiply by a quantity such as $x + 4$ because we do not know if it is positive or negative and therefore could not be sure as to the sense of the inequality.

We combine the terms on the left by finding a common denominator and forming a single fraction.

$$\frac{2x^2 - 5x - 3}{x + 4} > 0$$

or

$$\frac{(2x + 1)(x - 3)}{x + 4} > 0$$

$2x + 1 > 0$

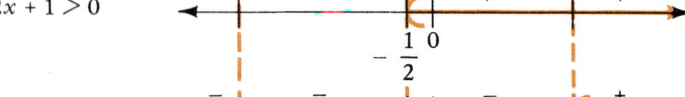

$x - 3 > 0$

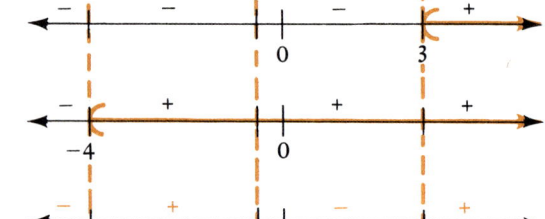

$x + 4 > 0$

The graph of the solution set is

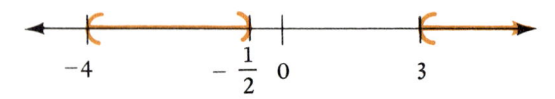

or in interval notation $\left(-4, -\dfrac{1}{2}\right) \cup (3, +\infty)$.

Exercise 2.6.1
Solve for x.

1. $x^2 - 5x + 6 < 0$
2. $x^2 + 3x - 10 < 0$
3. $x^2 - x - 12 > 0$
4. $x^2 + 5x + 4 \geqslant 0$
5. $2x^2 - 9x - 18 \leqslant 0$
6. $\dfrac{x + 4}{x - 1} > 0$
7. $\dfrac{x^2 - x - 2}{x - 4} < 0$
8. $\dfrac{x^2 - 1}{x - 5} > -1$

9. $\dfrac{x^2 - 11}{x - 3} \leq 7$ 10. $\dfrac{x^2 - 37}{x + 4} \geq -4$

11. $\dfrac{x - 5}{5 - x} \geq 0$ 12. $\dfrac{x^2 + 17}{x + 2} < \dfrac{11}{2}$

2.7 VERBAL PROBLEMS

Goals

Upon completion of this section you should be able to:

1. Outline and solve verbal problems.

* *

In sections throughout the text you will be asked to work verbal problems. The applications of the techniques of solving equations are a part of solving these problems, but the initial task is to arrive at an equation that satisfies the conditions within the problem. In this section we will present by example the basic methods of solving verbal problems.

The first step, and consequently the most important, in solving a verbal problem is to determine which unknown is the "key" to the other unknown quantities in the problem. Consider the following problem.

EXAMPLE

Bill is three years older than Karen, and Jim is seven years younger than Karen. If the sum of their ages is 86, how old is each?

Our thinking should proceed in this direction: "If we knew Karen's age, we could add three to it and find Bill's age. Also we could subtract seven from Karen's age and find Jim's age." This should give rise to the following outline:

$$x = \text{Karen's age}$$
$$x + 3 = \text{Bill's age}$$
$$x - 7 = \text{Jim's age}$$

If the sum of their ages is 86 we can write the equation

$$(x) + (x + 3) + (x - 7) = 86$$

Solving we obtain $x = 30$. Thus Karen is 30, Bill is 33, and Jim is 23.

Special note should be taken of the following facts which are true for any verbal problem:

1. You must first determine what x will represent before you can proceed to outline a problem.
2. The solution to the equation is *not* necessarily the solution to the problem. The answer must satisfy the conditions set forth in the problem, and the question asked must be completely answered. This means you must check your answer *both* in the equation and the verbal problem itself.

EXAMPLE

Find two positive integers that differ by 5 and have a product of 84. We let

$$x = \text{first integer}$$
$$x + 5 = \text{second integer}$$

Then $x(x + 5) = 84$ giving

$$x^2 + 5x = 84$$
$$x^2 + 5x - 84 = 0$$
$$(x + 12)(x - 7) = 0$$
$$x = -12 \quad \text{or} \quad x = 7$$

Notice that $x = -12$ and $x = 7$ both check in the original equation, *but* a condition of the problem is that x be positive. Therefore $x = -12$ is not a solution. Also note that we are asked to find both integers. The answer to the problem is therefore "The two integers are 7 and 12."

EXAMPLE

A person invests $8,000 in two securities. One pays an annual interest of 4% and the other 7%. The total interest on both securities for the first year was $455. How much was invested at each rate?

If we let $\quad x = $ the amount invested at 4%

then $\quad 8,000 - x = $ the amount invested at 7%

Now if x dollars is invested at 4%, then the interest for the year on this investment would be found by multiplying the investment x by the rate of interest. Thus $0.04x$ represents the interest on this investment. Similarly $0.07(8,000 - x)$ represents the interest on the other investment. The sum of the interests equals $455. Thus

$$0.04x + 0.07(8000 - x) = 455$$

Solving we obtain

$$x = \$3,500 \quad \text{and} \quad 8,000 - x = \$4,500$$

EXAMPLE

Two cars start at the same point and travel in opposite directions, one at 90 kilometers per hour and the other at 60 kilometers per hour. How many hours does it take them to get 450 kilometers apart?

In this problem we make use of the formula $d = rt$ (distance = rate × time). A diagram might be helpful to visualize the situation.

If we let t = the number of hours, then $90t$ must represent the distance traveled by the first car (distance = 90 × t) and $60t$ represents the distance traveled by the second car. Now, since the sum of the distances is 450 kilometers, we obtain the equation

$$90t + 60t = 450$$

Solving we obtain $t = 3$ hours

EXAMPLE

A 12-liter solution of water and alcohol is 20% alcohol. How many liters of alcohol should be added to bring the solution up to 25% alcohol?

Let $\quad x$ = number of liters of alcohol to be added

In a problem dealing with mixtures, concentrate only on one quantity. In this case either consider the water or alcohol but *not* both.

Let's consider the alcohol. How much alcohol do we start with? We have 12 liters of solution and 20% of it is alcohol. Therefore 0.20(12) represents the amount of pure alcohol.

We are adding x liters of pure alcohol, so $0.20(12) + x$ represents the total amount of alcohol we have in the final solution. Since we are obtaining a total of $12 + x$ liters of solution and it is 25% alcohol, then $0.25(12 + x)$ must also represent the amount of pure alcohol in the final solution. Thus

$$0.20(12) + x = 0.25(12 + x)$$

is our desired equation. Solving we obtain

$$x = \frac{60}{75} = \frac{4}{5} \text{ liters}$$

2.7 VERBAL PROBLEMS

Exercise 2.7.1

Outline and solve the following:

1. Find two consecutive positive even integers whose product is 840.

2. The hypotenuse of a right triangle is 10 centimeters long. If the sum of the two legs is 12 centimeters, find their lengths. (*Hint*: Use the Pythagorean Theorem.)

3. A rectangular plot of ground is to be made into a garden with a path of uniform width along all four sides. If the plot of ground is 15 meters by 30 meters and that portion inside the path contains 216 square meters, how wide is the path?

4. What number must be added to the numerator and denominator of $\frac{5}{8}$ to yield a fraction equal to $\frac{7}{8}$?

5. If part of $15,000 is invested at 6% and the remainder at 8% and the annual income from the total investment is $1,072, how much is invested at each rate?

6. If the width of a rectangle is 2 meters more than half the length and the perimeter is 64 meters, find the length and width.

7. If a man can row a boat 6 kilometers downstream and 6 kilometers back upstream in 4 hours and the rate of the current is 2 kilometers per hour, what is his rate of rowing in still water? The formula to be used is $d = rt$ (distance equals the product of the rate and time).

8. The sum of the reciprocals of two consecutive even integers is $\frac{13}{84}$. Find the integers.

9. One car leaves the city traveling north at an average rate of 45 kilometers per hour. If a second car leaves the same place 20 minutes later and averages 60 kilometers per hour, how long will it take the second car to overtake the first and at what distance from the city will it occur? (*Hint:* Draw a diagram.)

10. One pipe can fill a tank in 1 hour and another (the drain) can empty the tank in $1\frac{1}{2}$ hours. A worker opens the valve to fill the tank and 20 minutes later notices that the drain has been left open. If the worker then closes the drain, how long will it take the first pipe to finish filling the tank?

11. How much water will be required to dilute 15 liters of a 12% alcohol solution to obtain a 5% solution?

12. One worker can complete a certain job in 10 hours. A second worker can do the same job in 12 hours. How long will it take them to complete the job if they work together?

13. A woman invests part of $10,000 at 5% and the remainder at 8% per year. If her total interest for one year is $650, how much did she invest at each rate?

14. What negative number added to the square of one-half itself is equal to 24?

CHAPTER REVIEW

Solve for x.

1. $2x + 3 = 3(2x + 5)$
2. $\dfrac{x}{2} + \dfrac{2}{3}x = 7$
3. $\dfrac{3}{2x} = \dfrac{1}{x} + \dfrac{1}{2}$
4. $\dfrac{x}{x+7} = \dfrac{x-1}{x}$
5. $\dfrac{1}{x-2} + \dfrac{1}{x+2} = \dfrac{4}{x^2 - 4}$
6. $\dfrac{x+1}{x+2} - \dfrac{x-3}{x+5} = \dfrac{4}{x^2 + 7x + 10}$
7. $x^2 + 10x + 21 = 0$
8. $6x^2 + 7x = 3$
9. $\dfrac{2}{x-3} + 1 = \dfrac{6}{x^2 - 3x}$
10. $2x^2 + 8x + 7 = 0$

Compute $b^2 - 4ac$ and give the nature of the roots.

11. $x^2 - 3x - 18 = 0$
12. $x^2 + 6x + 9 = 0$
13. $2x^2 - 3x + 2 = 0$
14. $3x^2 + 5x + 2 = 0$

Perform the following operations:

15. $(3 + 4i) + (6 - 2i)$
16. $(7 - 2i) - (5 - 3i)$
17. $(2 - i)(3 + 8i)$
18. $(3 + 9i) \div (2 - 3i)$
19. $4 \div 2i$

Solve for x.

20. $x^2 + 1 = 0$
21. $x^2 - 2x + 5 = 0$
22. $x^2 - 5x + 7 = 0$
23. $x^3 - 125 = 0$
24. $x^4 + 3x^2 - 10 = 0$
25. $x^{\frac{1}{2}} - 7x^{\frac{1}{4}} + 10 = 0$

26. $x - 9x^{\frac{1}{2}} + 20 = 0$

27. $x^{\frac{2}{3}} + 3x^{\frac{1}{3}} - 4 = 0$

28. $4 + \sqrt{2x} = x$

29. $\sqrt{x + 15} - \sqrt{2x + 13} = \sqrt{x + 10}$

30. $|3x - 4| = 9$

31. $|x^2 + 3x - 6| = 4$

32. $|10x^2 - 13x + 1| = 4$

33. $2x + 5 \geq 6x - 7$

34. $\frac{1}{4}(x + 3) - 2x < \frac{2}{3}(2x - 3) + \frac{1}{2}x$

35. $|3x - 2| < 7$

36. $|6x + 5| \geq 11$

37. $x^2 + 4x - 21 < 0$

38. $3x^2 + 17x + 10 \geq 0$

39. $\frac{x - 5}{5 - x} \leq 0$

40. $\frac{x^2 - 13}{x - 3} \geq 3$

41. Find the angles of a triangle if the first exceeds the second by 20° and the third is 12° more than twice the second. (The sum of the angles of a triangle is 180°.)

42. Two positive numbers are such that one is twice the other. If the smaller number is increased by 2 and the larger is increased by 3, the product of the new numbers is 91. Find the original numbers.

PRACTICE TEST

1. Compute $b^2 - 4ac$ and give the nature of the roots.

    a. $x^2 + 3x - 2 = 0$ b. $6x^2 - x - 2 = 0$ c. $3x^2 - 4x + 5 = 0$

2. Perform the following operations:

    a. $(5 - i) - 3(2i)$ b. $(2 - 3i)(4 + i)$ c. $(3 + 7i) \div (1 - i)$

3. Solve for x.

    a. $3(x + 3) = x - 5$

    b. $\frac{x}{3} - \frac{2}{5}x + 1 = \frac{1}{3}x$

    c. $\frac{1}{x + 1} - \frac{1}{2} = \frac{1}{3x + 3}$

    d. $2x^2 + 3x = 2$

    e. $\frac{3}{x + 1} + \frac{3}{x^2 + x} = -2$

    f. $x^2 - 3x + 1 = 0$

    g. $x^2 - 2x + 2 = 0$

    h. $x^3 - 1 = 0$

    i. $x^4 - 10x^2 + 9 = 0$

    j. $\sqrt{3x + 1} + \sqrt{2x - 1} = 7$

k. $|x^2 + 5x + 7| = 2$ l. $\dfrac{2}{5}(x - 1) > \dfrac{1}{3}(2x + 5)$

m. $|5x - 1| \leq 4$ n. $\dfrac{x^2 + x - 6}{x - 1} > 0$

4. A customer buys x articles at $0.20 each and twice as many articles at $0.25 each. If the cost of all the articles is $4.90, find the number bought at each price.

PRACTICE TEST

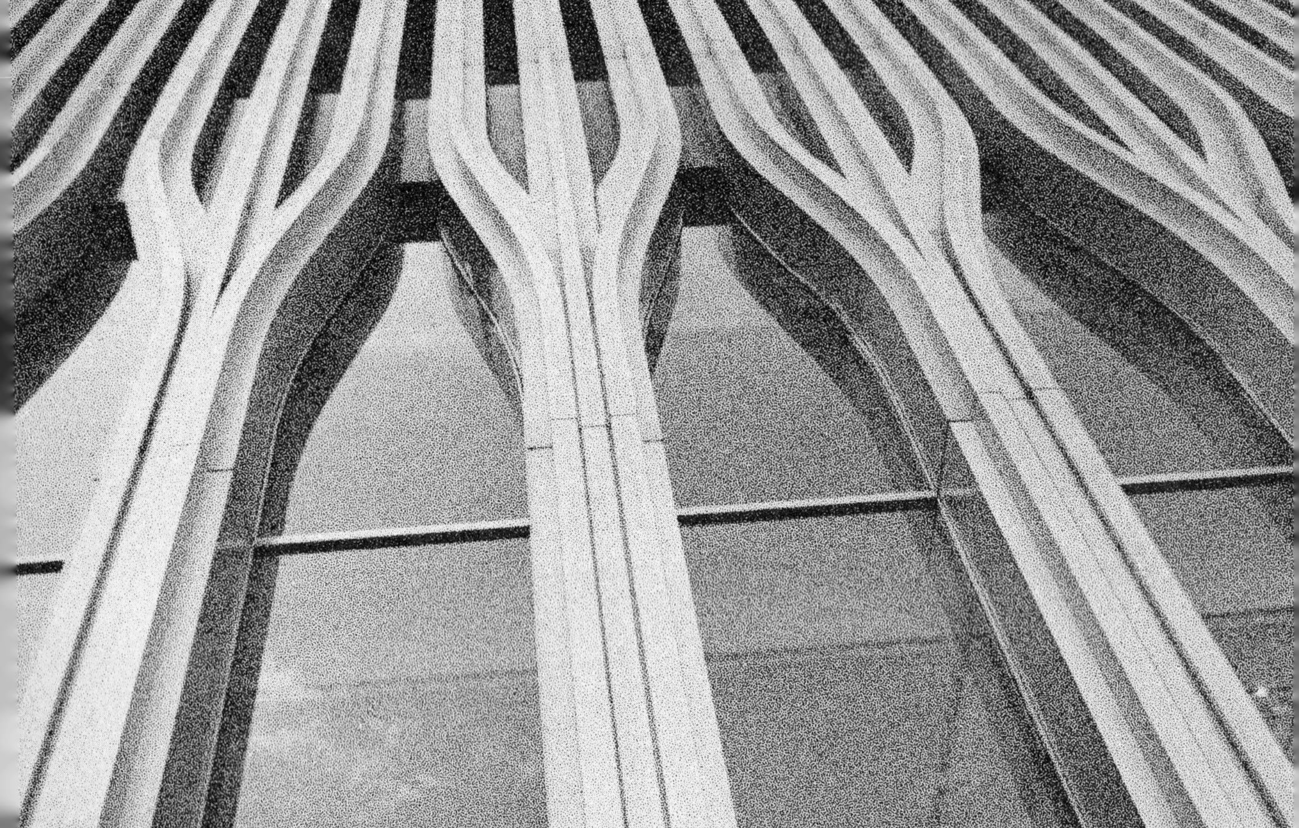

Relations and Functions

3

3

The idea of function may well be the most important concept in mathematics. Progress in the scientific fields would be greatly hampered without this concept. In this chapter we will consider only real functions and treat the concept in precise terms and investigate some of its mathematical implications.

3.1 DEFINITIONS

Goals

Upon completion of this section you should be able to:

1. Define a function.
2. Determine if a given relation is a function.
3. Find the domain and range of a given function.

* *

If we are given two sets of numbers, we can establish some rule that shows a relation between elements of the first set and elements of the second set.

For instance, given the two sets

$$A: \quad \{1, 3, 5\}$$
$$B: \quad \{2, 4, 6, 8, 10\}$$

the rule "Relate a number in set A to a number in set B so that the number in set B is twice as large as the number in set A" would give

1 is related to 2

3 is related to 6

5 is related to 10

We can express this more compactly as a set of ordered pairs

$$\{(1,2),(3,6),(5,10)\}$$

where the first number (*abscissa*) in each pair is from set A and the second number (*ordinate*) is from set B.

The rule "Relate a number in set A to a number in set B that is larger" would give

1 is related to 2, 4, 6, 8, and 10

3 is related to 4, 6, 8, and 10

5 is related to 6, 8, and 10

Writing this as a set of ordered pairs we have

$$\{(1,2),(1,4),(1,6),(1,8),(1,10),(3,4),(3,6),(3,8),(3,10),(5,6),(5,8),(5,10)\}$$

Definition   A *relation* is a set of ordered pairs.

The set of all first numbers in the ordered pairs of a relation is called the *domain.* In the first example given in this section the domain is $\{1,3,5\}$. It is also the domain of the second example.

The set of second numbers in the ordered pairs of a relation is called the *range.* In the first example the range is $\{2,6,10\}$. The range in the second example is $\{2,4,6,8,10\}$. Sometimes the numbers in the range are referred to as *images* and the range is called the *image set.*

Definition   A *function* is a relation such that each element of the domain has exactly one image.

From this definition we see that the first example is a function, whereas the second example is not a function.

In mathematics, the rule relating the elements of the domain to elements of the range is usually given as an algebraic expression. For instance, in the first example, if x represents an element of set A, then $2x$ represents the image of x in set B.

The notation $f(x)$ is read "f of x" or "f at x" and means "the image of x."

$f(x) = 2x$ means "the image of x is a number that is twice x"

$f(x) = x^2 + 1$ means "the image of x is a number that is one more than the square of x"

The notation is only used when the rule actually gives a relation that is a function.

3.1 DEFINITIONS

The domain of a real function can be the set of real numbers or any subset of the real numbers and is sometimes limited by the algebraic expression of the rule. For instance,

$$f(x) = \frac{1}{x}$$

has a domain of all real numbers except zero, i.e., $(-\infty, 0) \cup (0, +\infty)$. Also

$$f(x) = \sqrt{x}$$

has a domain of all nonnegative real numbers, i.e., $[0, +\infty)$.

If we are given an algebraic equation in two variables, it may be important to know if one variable is a function of the other.

EXAMPLE

"Is y a function of x in the equation $x^2 + 2x - y = 5$?" If so, determine the domain and range.

We know that y is a function of x if, for each value of x, there is exactly one value of y (image). Therefore, to answer the question, we must first solve the equation for y in terms of x and then determine if the definition of a function is satisfied. Thus

$$y = x^2 + 2x - 5$$

With the equation in this form we can see that any value for x will yield exactly one value of y. Thus the answer to the question is "yes."

To determine the domain and range, we need to think negatively. In other words "Is any real number *eliminated* from the domain or range for any reason?" Inspecting $y = x^2 + 2x - 5$ shows that any real number substituted for x will give a real value for y. No real number is eliminated as a possible value of x, so the domain is the set of all real numbers, i.e., $(-\infty, +\infty)$.

To determine the range we will solve the equation for x in terms of y by use of the quadratic formula. This gives

$$x = -1 \pm \sqrt{6 + y}$$

which implies that $6 + y \geq 0$ is a condition imposed on y so that x will be real. Thus the range is the set of all reals y such that $y \geq -6$, which in interval notation is $[-6, +\infty)$.

EXAMPLE

"Is y a function of x in the expression $y^2 = x - 1$?" First we solve for y obtaining

$$y = \pm \sqrt{x - 1}$$

We see that a value of x will yield *two* values of y. Therefore y is not a function of x.

Exercise 3.1.1

In each of the following relations consider (from the set of real numbers) the replacement set for x as the domain and the replacement set for y as the range.

a. Find the domain.

b. Find the range.

c. Is the relation a function?

1. $y = x^2 - 1$
2. $\dfrac{1}{y} = x$
3. $y^2 = x + 3$
4. $y = x^3$
5. $x = y^3$
6. $x^2 + y^2 = 9$
7. $y = \sqrt{9 - x^2}$
8. $y = \dfrac{1}{x - 1}$
9. $x = \sqrt{y^2}$
10. $y = |x^2 + 3|$
11. $y = \dfrac{x + 2}{x - 3}$
12. $y = \sqrt{x^2 - 9}$
13. $y = \dfrac{|x - 2|}{x}$
14. $x^2 + y^2 < 9$
15. $y = \dfrac{\sqrt{x^2}}{x}$
16. $y = \dfrac{1}{\sqrt{x - 2}}$

3.2 GRAPHS OF RELATIONS

Goals

Upon completion of this section you should be able to:

1. Find the length of a line segment on a line parallel to either axis.
2. Use the distance formula to find the distance between any two points in the plane.
3. Use the midpoint formula.
4. Construct a "table of values" and graph an algebraic relation.

* *

One method of uniquely naming each point in the plane is the Cartesian coordinate system (so named for the French mathematician René Descartes, 1596–1650).

This rectangular coordinate system is constructed with two perpendicular number lines in the plane that intersect at zero on each line. One line is vertical and the other is horizontal, with positive direction upward and to the right, respectively. The two lines are called *coordinate axes,* and their point of intersection is called the *origin.* The horizontal axis refers to the domain and is usually referred to as the *x-axis*, and the vertical axis refers to the range and is referred to as the *f(x)-axis* or *y-axis*.

The four regions of the plane formed by the *x-* and *y*-axes are called *quadrants* and are numbered in a counterclockwise direction starting with the upper right.

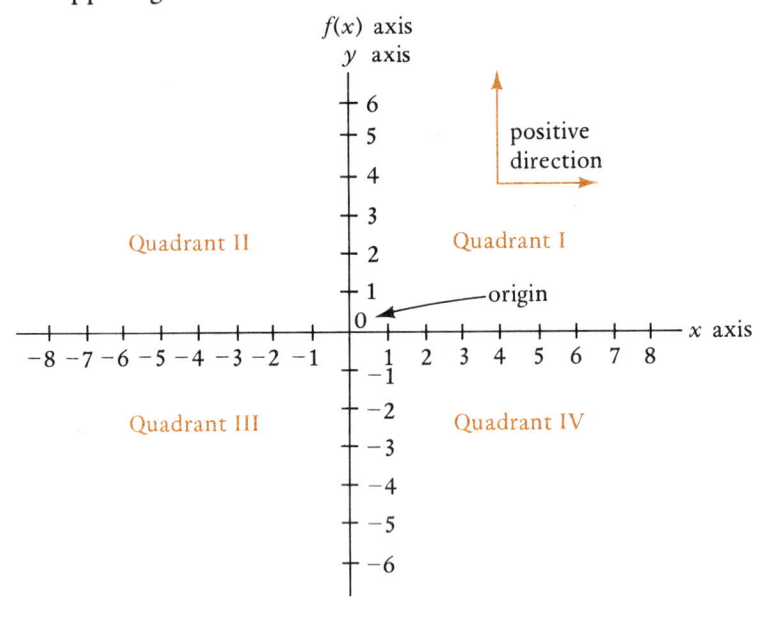

Any point in the plane can be located by moving first along the *x*-axis and then parallel to the *y*-axis. If we agree to write a pair of numbers such as (5,3) so that the first number represents the distance and direction from the origin along the *x*-axis and the second number represents the distance and direction parallel to the *y*-axis, then this ordered pair of numbers represents one and only one point on the plane. In the Cartesian coordinate system, points on the plane are always represented by an ordered pair $[x, f(x)]$ or (x,y). The real numbers x and $f(x)$ or y of this ordered pair are called the *coordinates* of the point.

The length of a line segment on the coordinate plane, or the distance from point P_1 to point P_2 is a useful tool in many instances.

When a line segment is parallel to one of the axes, its length is easily found by one of the following theorems.

Theorem   The length of the line segment P_1P_2 where P_1 has coordinates (x_1, y_1) and P_2 has coordinates (x_2, y_1) is $|x_1 - x_2|$.

Theorem   The length of the line segment P_1P_2 where P_1 has coordinates (x_1, y_1) and P_2 has coordinates (x_1, y_2) is $|y_1 - y_2|$.

Do you see why these theorems are true? Why is the absolute value used?

The length of a line segment that is not parallel to an axis is not quite so obvious, and we will here develop the formula for its length.

Our problem is to determine the distance between P_1 with coordinates (x_1, y_1) and P_2 with coordinates (x_2, y_2), i.e., any two points on the plane.

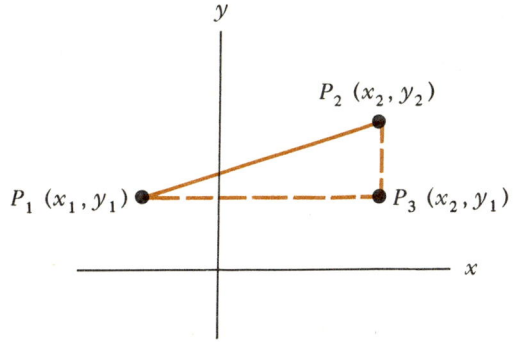

We first construct a line through P_1 parallel to the x-axis and a line through P_2 parallel to the y-axis. These lines will be perpendicular at a point P_3 which will have coordinates (x_2, y_1). Why? We now have a right triangle $P_1P_3P_2$ with hypotenuse P_1P_2.

Using the Pythagorean Theorem from plane geometry gives

$$\overline{P_1P_2}^2 = |x_1 - x_2|^2 + |y_1 - y_2|^2$$
$$\overline{P_1P_2}^2 = (x_1 - x_2)^2 + (y_1 - y_2)^2$$

Why can we drop the absolute value symbol in this step? We now take the square root of each side obtaining

$$\overline{P_1P_2} = \sqrt{(x_1 - x_2)^2 + (y_1 - y_2)^2}$$

3.2 GRAPHS OF RELATIONS

We only use the positive square root since distance is a nonnegative quantity. This last form is known as the *distance formula* and can be used to find the distance between any two points on the plane.

EXAMPLE

Find the distance from P_1 (3,7) to P_2 (−4,6).

$$\overline{P_1 P_2} = \sqrt{[3-(-4)]^2 + (7-6)^2}$$
$$= \sqrt{49 + 1}$$
$$= \sqrt{50}$$
$$= 5\sqrt{2}$$

Another formula that is used is the *midpoint formula*. The midpoint of a line segment $P_1 P_2$ would be a point midway between P_1 and P_2. The coordinates of this point would therefore be the averages of the coordinates of the endpoints. Thus, the following theorem.

Theorem

The midpoint of the line segment having endpoints $P_1(x_1, y_1)$ and $P_2(x_2, y_2)$ has coordinates

$$\left(\frac{x_1 + x_2}{2}, \frac{y_1 + y_2}{2} \right)$$

EXAMPLE

Find the midpoint of the line segment joining the points (5, −7) and (3,1).

The midpoint is

$$\left(\frac{5+3}{2}, \frac{-7+1}{2} \right) = (4,-3)$$

You may want to convince yourself that (4,−3) is the midpoint by finding its distance from each endpoint of the line segment.

The graph of a relation between x and y can be represented on the coordinate system by expressing this relation as ordered pairs and locating the points on the plane represented by these ordered pairs. It is generally not possible to locate all points given by an algebraic equation, so we content ourselves with locating a sufficient number of points to establish a pattern, and then sketch the graph.

EXAMPLE

Sketch the graph of $y = 3x - 1$.

Our first task is to find ordered pairs (x,y) that are solutions to the equation. We accomplish this by assigning arbitrary values to x and finding the corresponding values of y. First, we note that the domain is the set of all real numbers. We can therefore assign any real number for x. For instance, if $x = 1$, then $y = 3(1) - 1 = 2$, so the pair $(1,2)$ is a solution to the equation and hence a point on the graph.

It is convenient to place these pairs in a "table of values." We will let x take the values $-1, 0, 1, 2$. We find the corresponding values of y and set up the table as follows.

x	-1	0	1	2
y	-4	-1	2	5

We now locate the points $(-1,-4)$, $(0,-1)$, $(1,2)$, and $(2,5)$ on the coordinate plane. Since the domain consists of all real numbers and the range also consists of all real numbers, there will be infinitely many points on the graph. We can see from these four points, however, that the graph appears to be a straight line. We may thus connect these points with a line.

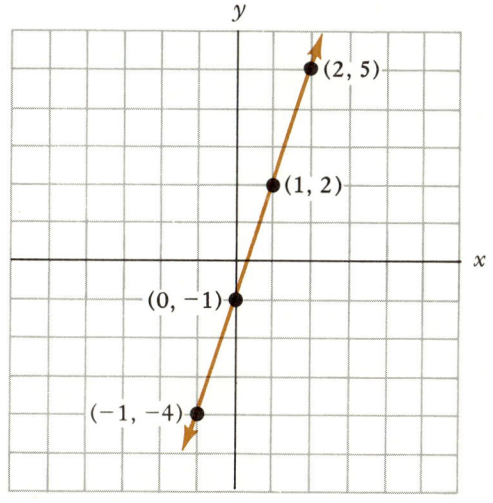

Graph of $y = 3x - 1$

These points establish a pattern, and we see that the graph of $y = 3x - 1$ is a straight line. It can be shown that the graph of any first-degree polynomial equation in two variables will form a straight line.

3.2 GRAPHS OF RELATIONS

EXAMPLE

Sketch the graph of $f(x) = x^2 - 3x - 4$. We again set up a table of values and plot enough points to establish a pattern.

x	−2	−1	0	1	2	3	4	5
$f(x)$	6	0	−4	−6	−6	−4	0	6

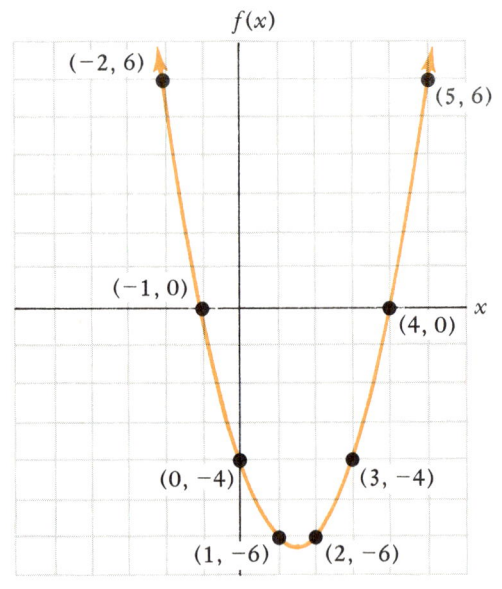

Graph of $f(x) = x^2 - 3x - 4$

The resulting curve is called a *parabola* and will be studied in more detail in the next section.

EXAMPLE

Sketch the graph of $f(x) = x^3 + x^2 - 6x$.

x	−4	−3	−2	−1	0	1	2	3
$f(x)$	−24	0	8	6	0	−4	0	18

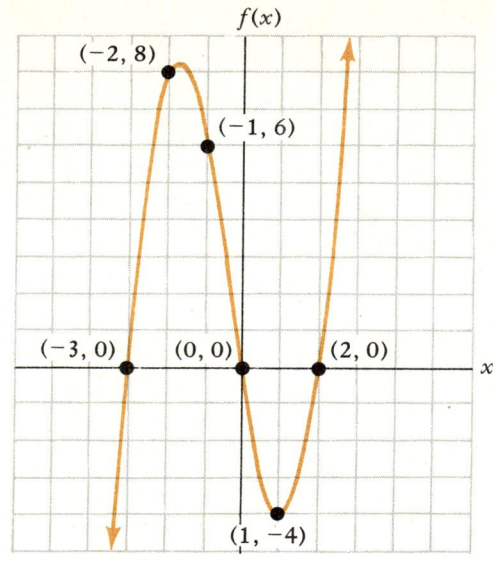

Graph of $f(x) = x^3 + x^2 - 6x$

The domain and range are each all real numbers. Is this a function?

EXAMPLE Sketch the graph of $y = |x + 3|$.

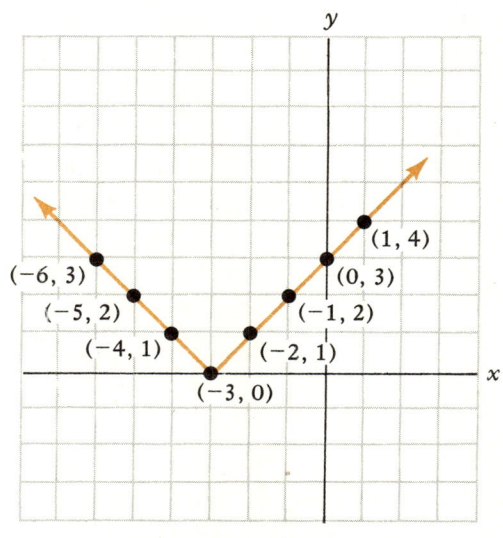

Graph of $y = |x + 3|$

3.2 GRAPHS OF RELATIONS

x	-6	-5	-4	-3	-2	-1	0	1
y	3	2	1	0	1	2	3	4

Notice here that the domain is all reals and the range is $[0,+\infty)$. Is this a function?

EXAMPLE

Sketch the graph of $y = \sqrt{x-2}$.

We first should notice that the domain is $[2,+\infty)$ and the range is $[0,+\infty)$. Taking some arbitrary values of x in this domain, we obtain the following table and graph.

x	2	3	4	5	6	11
y	0	1	$\sqrt{2}$	$\sqrt{3}$	2	3

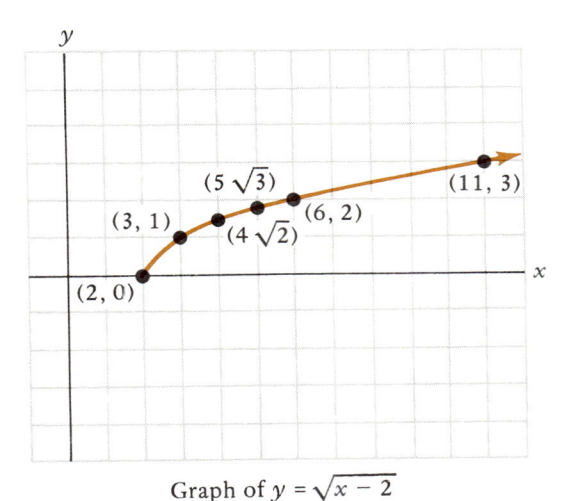

Graph of $y = \sqrt{x-2}$

Is this a function?

Exercise 3.2.1

1. Find the distance between the following pairs of points:

   a. $(0,-3), (0,5)$

   b. $(-2,8), (6,0)$

c. $(3,-1), (5,2)$ d. $(5,-2), (-1,-3)$

2. Show that the points $P_1(-2,1)$, $P_2(3,-4)$, and $P_3(5,-2)$ are the vertices of a right triangle.

3. Show that the points $A(-4,5)$, $B(0,1)$, and $C(3,-2)$ are collinear (i.e., lie on the same straight line).

4. Find the midpoint of each of the following line segments (the endpoints are given).

   a. $(3,-1), (5,7)$ b. $(4,5), (-10,7)$

   c. $(-5,1), (9,4)$ d. $(6,-13), (-1,8)$

5. Line segment AB has midpoint $(-1,6)$ and the coordinates of A are $(3,-4)$. Find the coordinates of B.

6. Give the domain and range of each relation and state whether the relation is a function. Graph the relation.

   a. $y = 2x - 1$ b. $f(x) = 6 - 3x$ c. $y = 5$

   d. $x = -3$ e. $x^2 + y^2 = 49$ f. $y = x^2 + 4x - 1$

   g. $f(x) = 2x^2 - 3$ h. $y = |x + 5|$ i. $f(x) = \sqrt{x + 1}$

   j. $y = x^3 - 3x^2 + 1$ k. $x^2 + y^2 - 6x + 10y = -18$

   l. $f(x) = |5 - 2x|$ m. $y = \sqrt{x^2 - 3}$

3.3 THE CONIC SECTIONS

Goals

Upon completion of this section you should:

1. Know the geometric definition of the conic sections.
2. Know the standard form of the equations of the conic sections.
3. Be able to change a second-degree equation in two variables to the standard form of one of the conic sections and sketch its graph.
4. Be able to write the equation of a particular conic section, given the necessary information.

* *

The curves formed by second-degree equations in two variables are the *circle, ellipse, hyperbola,* and *parabola.* These curves are studied in detail in the subject of analytic geometry, but because their equations are also important in algebra, we will briefly discuss them in this section.

The curves are called *conic sections* or *conics* because they result from the intersection of a plane and a right circular cone, as is illustrated.

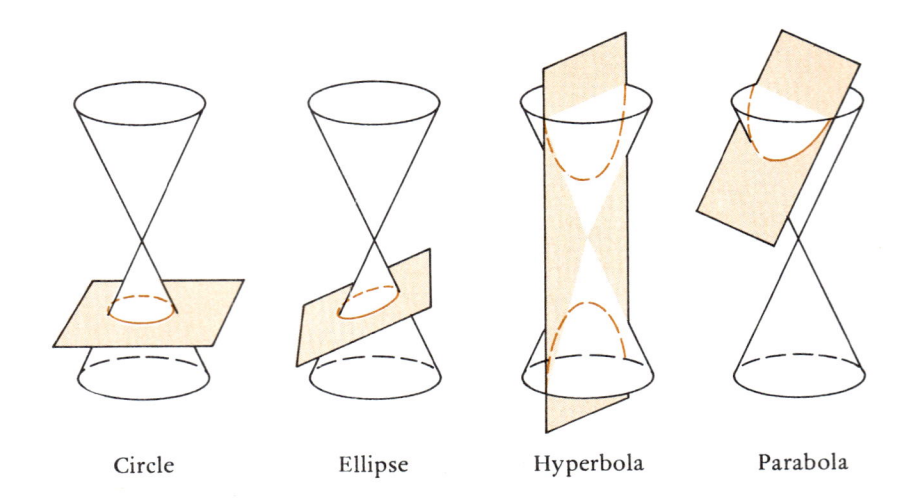

      Circle   Ellipse   Hyperbola   Parabola

We will first consider the circle.

Definition   A *circle* is the set of all points in a plane that are a given distance from a fixed point. The given distance is the length of any line segment from the fixed point (called the center) to a point on the circle. All of these line segments are called radii (singular radius). Often r is used to represent the radius, and the center is called (h,k).

Using this definition and letting (x,y) represent any point in the set so described, the distance formula gives

$$r = \sqrt{(x-h)^2 + (y-k)^2}$$

or

$$(x-h)^2 + (y-k)^2 = r^2$$

This is the *standard form* of the equation of a circle having center (h,k) and radius r.

The graph of a circle is easy to sketch when we know its center and radius.

EXAMPLE

Sketch the graph of a circle with center (2,−1) and a radius of 3. Write its equation in standard form.

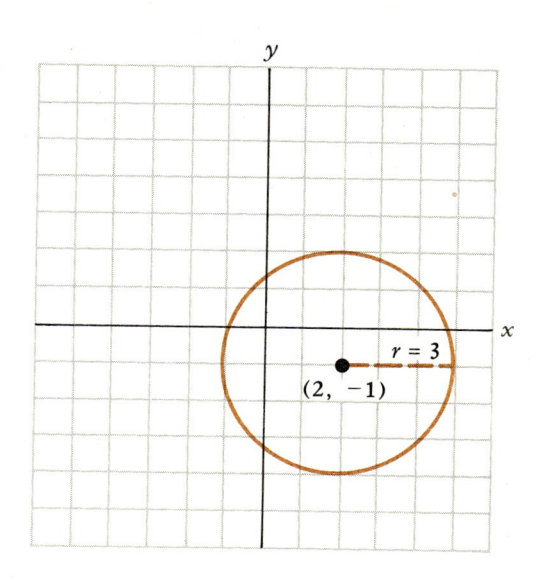

The equation in standard form is

$$(x - 2)^2 + [y - (-1)]^2 = 3^2$$

or

$$(x - 2)^2 + (y + 1)^2 = 9$$

EXAMPLE

$x^2 + y^2 + 4x + 8y = -4$ is the equation of a circle. Find its center and radius, and sketch its graph.

First we wish to put the equation in standard form. To do this we complete the square on each part by adding the correct number to both sides of the equation.

$$x^2 + 4x + 4 + y^2 + 8y + 16 = -4 + 4 + 16$$

or

$$(x + 2)^2 + (y + 4)^2 = 16$$

In this form we see that the center is (−2,−4) and the radius is 4.

3.3 THE CONIC SECTIONS

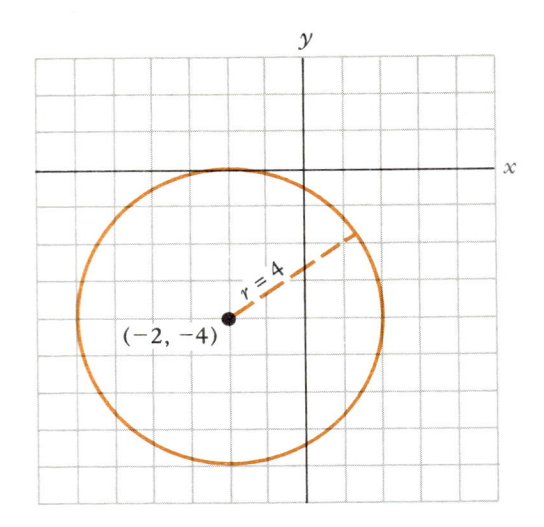

The domain is [−6,2] and the range is [−8,0]. Is this a function?

Exercise 3.3.1

1. Find the equation of a circle in standard form with center (−2,3) and radius 6.

2. Find the equation of a circle in standard form with center (0,0) and radius 7.

3. Each of the following is the equation of a circle. Put the equation in standard form and give the center and radius. Sketch the graph.

   a. $x^2 + y^2 + 2x - 12y = -28$ b. $x^2 + y^2 - 6x - 4y = 36$

   c. $x^2 + y^2 + 2x + 10y = -25$ d. $4x^2 + 4y^2 - 4x + 32y = 35$

4. Find the equation of a circle in standard form if the center is (3,−4) and the circle passes through the origin.

5. Find the equation of a circle in standard form if the endpoints of the diameter are (3,−9) and (−5,13). (*Hint*: Use the midpoint formula to determine the center.)

Definition

An *ellipse* is the set of all points in a plane, each such that the sum of its distances from two fixed points, called *foci* (singular focus), is a constant.

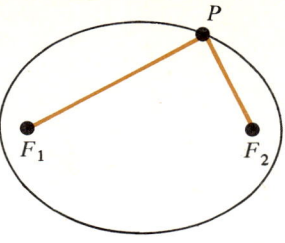

$\overline{PF}_1 + \overline{PF}_2$ is a constant

To obtain the equation of an ellipse, we place the ellipse on the coordinate plane so that the center of the ellipse is at (h,k).

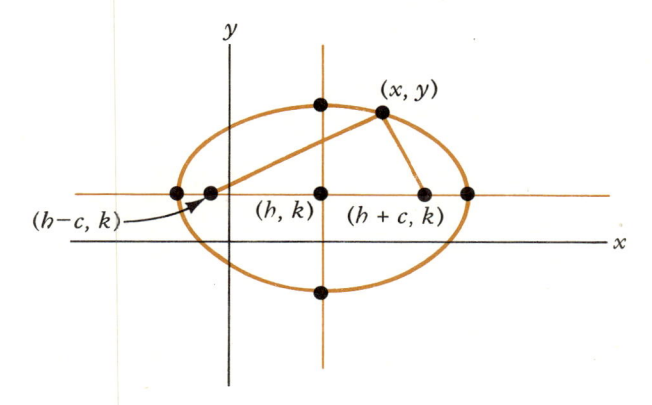

We shall call the distance from the center to the foci "c," thus the coordinates of the foci are $(h - c, k)$ and $(h + c, k)$. If we now pick any point on the ellipse and apply the definition, we see that the sum of its distances to the two foci must be a constant. We shall call this constant $2a$. The definition thus states

$$\sqrt{[x - (h - c)]^2 + (y - k)^2} + \sqrt{[x - (h + c)]^2 + (y - k)^2} = 2a$$

It is left as an exercise to show that this equation becomes

$$\frac{(x - h)^2}{a^2} + \frac{(y - k)^2}{a^2 - c^2} = 1$$

If we let $a^2 - c^2 = b^2$, then the equation can be stated as

$$\frac{(x - h)^2}{a^2} + \frac{(y - k)^2}{b^2} = 1$$

This is the *standard form* of the equation of an ellipse whose center is (h,k). $2a$ is the length of the horizontal axis, and $2b$ is the length

3.3 THE CONIC SECTIONS

of the vertical axis. If $a > b$, the ellipse is called a *horizontal* ellipse because its foci are on the horizontal axis. In this case, $a^2 - c^2 = b^2$. If $a < b$, we have a *vertical* ellipse, and the foci are on the vertical axis. In this case, $a^2 + c^2 = b^2$. The reason for this is left as an exercise.

EXAMPLE Sketch the graph of $\frac{(x-2)^2}{9} + \frac{(y+1)^2}{25} = 1$.

We immediately note that the center is $(2,-1)$, $a = 3$, and $b = 5$. First we plot the center. Then we move horizontally 3 units to the left and 3 units to the right of the center to find the endpoints of the horizontal axis. We then move vertically 5 units up and 5 units down from center to establish the endpoints of the vertical axis. These endpoints are sometimes referred to as *vertices*. We note that since $a < b$, we have a vertical ellipse and can now easily sketch the graph.

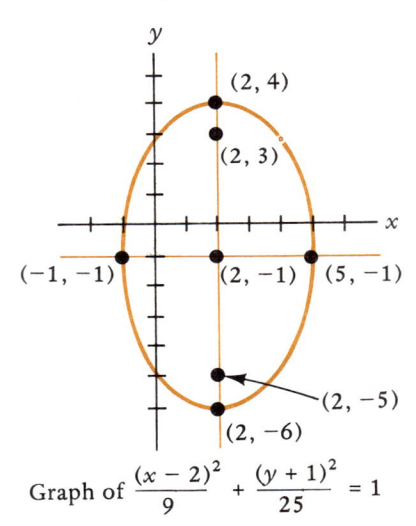

Graph of $\frac{(x-2)^2}{9} + \frac{(y+1)^2}{25} = 1$

Notice that it is not necessary to know the foci in order to sketch the graph of the ellipse. However, we know that the foci are a distance c from the center. Since, for a vertical ellipse

$$b^2 - c^2 = a^2$$

or

$$c^2 = b^2 - a^2$$

and since $a = 3$ and $b = 5$, then

$$c^2 = 25 - 9$$

or

$$c = 4$$

Thus the foci are (2,3) and (2,−5). The domain of the relation is [−1,5], and the range is [−6,4].

EXAMPLE $9x^2 + 18x + 16y^2 - 64y = 71$ is the equation of an ellipse. Find its center, foci, and vertices (endpoints of the axes). Sketch the graph.

We first wish to get the equation in standard form. To do this we factor 9 from the x terms and 16 from the y terms.

$$9(x^2 + 2x + \_\_) + 16(y^2 - 4y + \_\_) = 71$$

We next complete the squares.

$$9(x^2 + 2x + 1) + 16(y^2 - 4y + 4) = 71 + 9 + 64$$
$$9(x + 1)^2 + 16(y - 2)^2 = 144$$

or

$$\frac{(x + 1)^2}{16} + \frac{(y - 2)^2}{9} = 1$$

The equation is now in standard form, and we see that the center is (−1,2). We also note that it is a horizontal ellipse since $a = 4$ is greater than $b = 3$. Since it is a horizontal ellipse

$$a^2 - c^2 = b^2$$

or

$$c^2 = a^2 - b^2$$
$$= 16 - 9$$
$$= 7$$

Thus

$$c = \sqrt{7}$$

and the coordinates of the foci are $(-1 - \sqrt{7}, 2)$ and $(-1 + \sqrt{7}, 2)$. The vertices of the horizontal axis are (−5,2) and (3,2), and the vertices of the vertical axis are (−1,−1) and (−1,5).

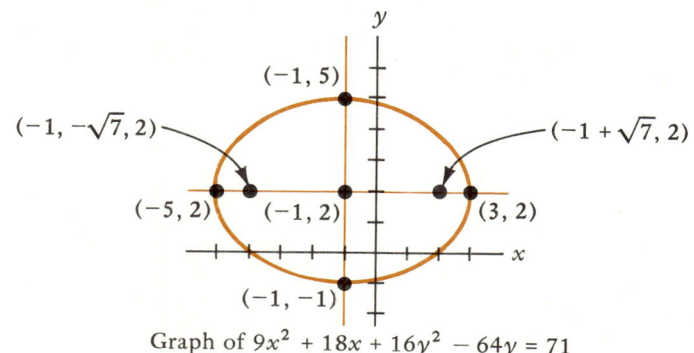

Graph of $9x^2 + 18x + 16y^2 - 64y = 71$

State the domain and range of this relation.

3.3 THE CONIC SECTIONS

Exercise 3.3.2

1. Find the center, vertices, and foci of the following. Graph each ellipse.

   a. $\dfrac{(x-3)^2}{25} + \dfrac{(y-1)^2}{9} = 1$

   b. $\dfrac{(x+1)^2}{16} + \dfrac{(y-4)^2}{36} = 1$

   c. $\dfrac{x^2}{9} + \dfrac{y^2}{25} = 1$

   d. $4x^2 + 16y^2 - 16x + 160y = -352$

2. Write the equation of the ellipse whose vertices are $(-10,2)$, $(7,2)$, $(-3,-1)$, and $(-3,5)$.

3. Write an equation that describes the set of all points in a plane satisfying the property that the sum of the distances from each to the points $(-4,3)$ and $(2,3)$ is 10.

4. Show that the equation

   $$\sqrt{[x-(h-c)]^2 + (y-k)^2} + \sqrt{[x-(h+c)]^2 + (y-k)^2} = 2a$$

   becomes $\dfrac{(x-h)^2}{a^2} + \dfrac{(y-k)^2}{b^2} = 1$ if we substitute b^2 for $a^2 - c^2$.

5. Suppose we have a vertical ellipse as shown in the diagram. The definition of the ellipse tells us that $\overline{PF}_1 + \overline{PF}_2$ is a constant which we may arbitrarily call $2b$. Thus $\overline{PF}_1 + \overline{PF}_2 = 2b$. Using the distance formula and letting $b^2 = a^2 + c^2$, derive the standard form of the equation of the ellipse.

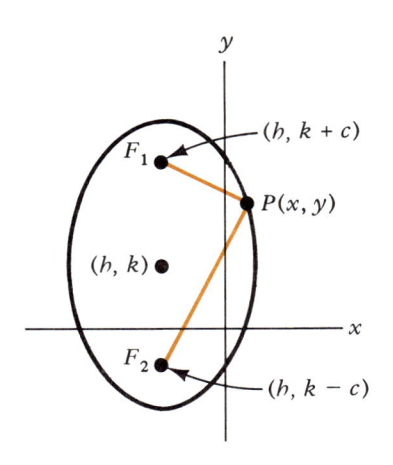

86 RELATIONS AND FUNCTIONS

Definition A *hyperbola* is the set of all points in a plane, each such that the absolute value of the difference of its distances to two fixed points (foci) is a constant.

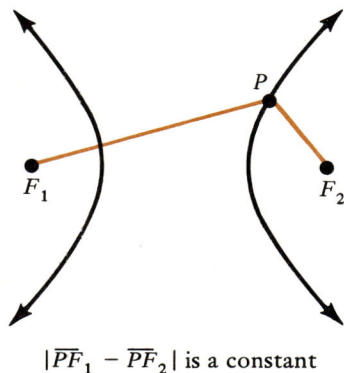

$|\overline{PF}_1 - \overline{PF}_2|$ is a constant

Using methods similar to those for the ellipse, we can establish the *standard form* for the equation of the hyperbola. This form is

$$\frac{(x-h)^2}{a^2} - \frac{(y-k)^2}{b^2} = \pm 1$$

where (h,k) is the *center*. The foci are a distance c from the center, and $a^2 + b^2 = c^2$. If the right side of the equation is $+1$, we have a *horizontal* hyperbola, and if the right side of the equation is -1, the hyperbola is *vertical*. The hyperbola is one of the easiest curves to sketch. Using (h,k) as the center, we construct a rectangle that is $2a$ units wide and $2b$ units high. We then draw and extend the diagonals of the rectangle. These extended diagonals are the *asymptotes* for the branches of the hyperbola. The *asymptote* has the characteristic that as we move out on any branch of the hyperbola, the curve continually approaches the asymptote but never reaches it. The hyperbola can then be easily sketched.

If the equation is

$$\frac{(x-h)^2}{a^2} - \frac{(y-k)^2}{b^2} = 1$$

the graph would be a horizontal hyperbola, as follows.

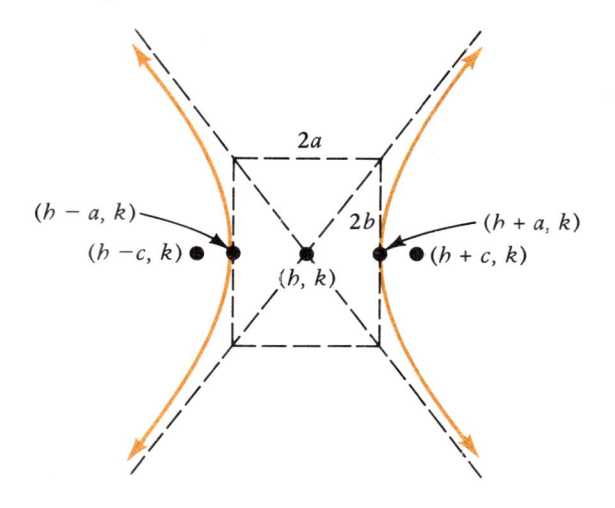

If the equation is

$$\frac{(x-h)^2}{a^2} - \frac{(y-k)^2}{b^2} = -1$$

the graph would be a vertical hyperbola.

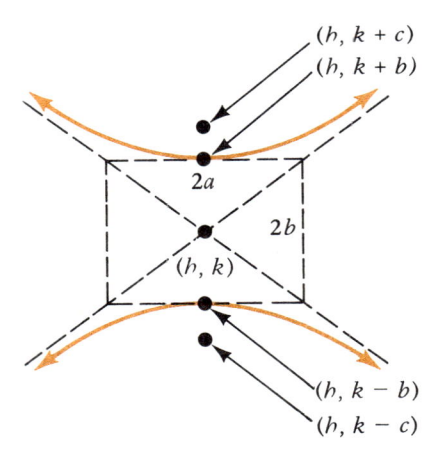

| EXAMPLE | Sketch the graph of $\frac{(x-2)^2}{9} - \frac{(y+1)^2}{16} = 1$.

We first note that it is a horizontal hyperbola. (Why?) We next locate the center at $(2, -1)$. Using this as the center, we construct a rectangle

88 RELATIONS AND FUNCTIONS

6 units (2a) long and 8 units (2b) high. We extend the diagonals of the rectangle to form the asymptotes for the hyperbola. Using (−1,−1) and (5,−1) as the vertices of the hyperbola, we may now sketch the graph.

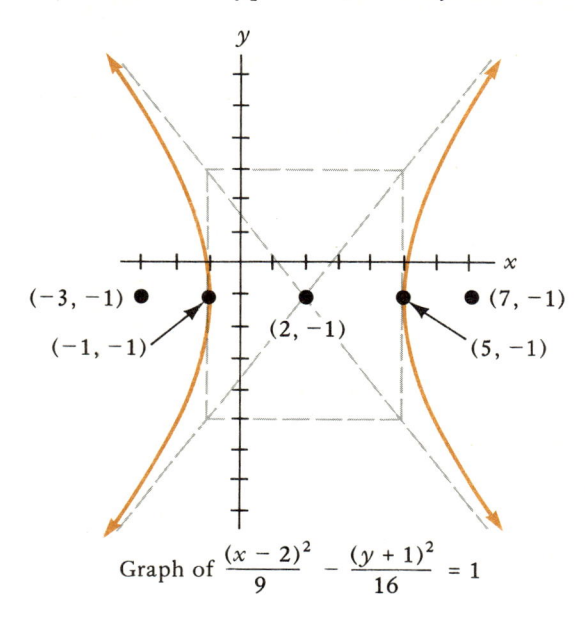

Graph of $\dfrac{(x-2)^2}{9} - \dfrac{(y+1)^2}{16} = 1$

The foci of the hyperbola are (−3,−1) and (7,−1). (−1,−1) and (5,−1) are the vertices of the hyperbola. What is the domain of this relation? What is the range?

Exercise 3.3.3

1. For each of the following hyperbolas, find the center, vertices, and foci. Draw the asymptotes and sketch the graph.

   a. $\dfrac{(x-2)^2}{9} - \dfrac{(y-3)^2}{16} = 1$ b. $\dfrac{(x+1)^2}{4} - \dfrac{(y-2)^2}{16} = -1$

   c. $y^2 - 4x^2 = 16$ (*Hint*: Put in standard form.)
   d. $9x^2 - 4y^2 - 54x - 32y = 19$

2. Write the equation of the hyperbola whose vertices are (1,−1) and (1,−9) and whose foci are (1,0) and (1,−10).

3. Write an equation that describes the set of all points in a plane for which the difference of the distances from each point to the points (−3,2) and (5,2) is 6.

3.3 THE CONIC SECTIONS 89

4. Using the diagram below along with the distance formula and the definition, establish the standard form for the equation of a horizontal hyperbola. Let $b^2 = c^2 - a^2$.

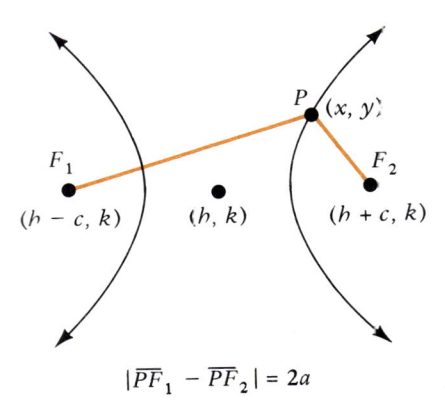

$|\overline{PF}_1 - \overline{PF}_2| = 2a$

5. Using the same method as in Problem 4, establish the standard form for the equation of a vertical hyperbola. (Use $|\overline{PF}_1 - \overline{PF}_2| = 2b$ and $a^2 + b^2 = c^2$.)

Definition A *parabola* is the set of all points in a plane equidistant from a fixed point (*focus*) and a fixed line (*directrix*) not containing the fixed point.

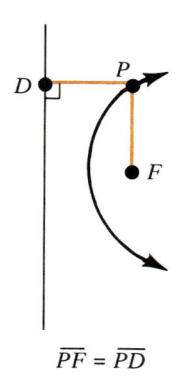

$\overline{PF} = \overline{PD}$

It should be mentioned immediately that a parabola is *not* one branch of a hyperbola as might be thought if they are carelessly drawn. They have different definitions, and a parabola does not have any asymptotes.

Using methods similar to those discussed earlier, we can establish the *standard form* for the equation of the parabola. The equation for a *vertical* parabola is

$$(x - h)^2 = 4p(y - k)$$

where (h,k) is the vertex of the parabola and the focus is $(h, k + p)$ and the directrix is $y = k - p$. The parabola opens upward if $p > 0$ and downward if $p < 0$.

EXAMPLE

Sketch the graph of $(x - 4)^2 = 8(y - 3)$.

This is a vertical parabola with vertex $(4,3)$, $p = 2$, and therefore the parabola opens upward. (Why?) The focus is $(4,5)$ and the directrix is $y = 1$.

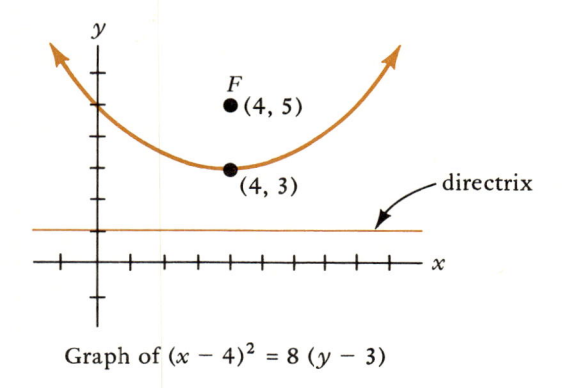

Graph of $(x - 4)^2 = 8(y - 3)$

The equation for a *horizontal* parabola is

$$(y - k)^2 = 4p(x - h)$$

where (h,k) is the vertex. The focus is $(h + p, k)$ and the directrix is $x = h - p$. The parabola opens to the right if $p > 0$ and to the left if $p < 0$.

EXAMPLE

Sketch the graph of $(y + 1)^2 = -12(x - 3)$.

This is a horizontal parabola with vertex $(3,-1)$, $p = -3$. The parabola opens to the left. (Why?) The focus is $(0,-1)$ and the directrix is $x = 6$.

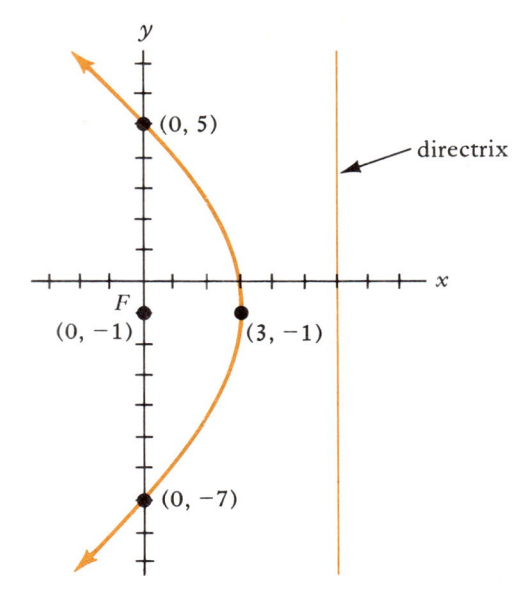

Graph of $(y + 1)^2 = -12(x - 3)$

Exercise 3.3.4

1. For each of the following parabolas, find the vertex, focus, and directrix. Sketch the graph.

   a. $(x - 3)^2 = 12(y + 1)$ b. $(y - 2)^2 = 4(x + 3)$

   c. $(x + 1)^2 = -8(y - 3)$ d. $y^2 - 2y + 8x - 15 = 0$

2. Write the equation of a parabola whose focus is $(5, -1)$ and whose directrix is $y = 3$.

3. Write an equation that describes the set of all points in a plane that are equidistant from the point $(3, 4)$ and the line $y = -2$.

4. Using the diagram that follows, along with the distance formula and the definition, establish the standard form of a vertical parabola.

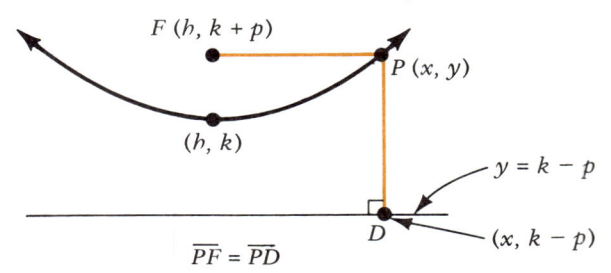

5. Using the same method as in Problem 4, establish the standard form of the equation of a horizontal parabola.

6. Put each of the following equations in standard form. Identify the conic and give important data such as center, vertices, foci, directrix, etc. Sketch the graph.

   a. $x^2 + 9y^2 - 10x + 36y + 52 = 0$ b. $x^2 - 8x - 8y + 8 = 0$
   c. $x^2 + y^2 + 8x - 6y - 5 = 0$ d. $9x^2 - 16y^2 + 72x + 96y + 144 = 0$
   e. $9x^2 + 4y^2 + 36x + 32y + 64 = 0$
   f. $y^2 - 12x + 8y + 40 = 0$ g. $x^2 + 25y^2 - 2x - 150y + 201 = 0$
   h. $x^2 - 12x - 4y + 48 = 0$ i. $x^2 + y^2 + 6x - 8y - 9 = 0$
   j. $9x^2 - 4y^2 + 36x + 40y - 100 = 0$
   k. $4x^2 - y^2 - 24x + 4y + 36 = 0$ l. $4x^2 + y^2 - 8x - 8y + 4 = 0$

3.4 LINEAR FUNCTIONS

Goals

Upon completion of this section you should:

1. Know the definition of the slope of a line.
2. Know the standard form, two-point form, and slope-intercept form of a straight line.
3. Be able to write the equation of a straight line in a specified form when given the necessary information.
4. Be able to find the slope and sketch the graph of a line from a given linear equation.

* *

The graph of a straight line parallel to the *y*-axis does not represent a function since each *x* value has many *y* values. All other straight lines represent functions. Remember that the graph of any first-degree equation in two variables will form a straight line.

We will call

$$ax + by = c$$

the *standard form* of the equation of a straight line. An important concept, related to the equation of a straight line, is that of *slope*. Intuitively we think of the slope of a line as the "steepness" of the line. The following definition gives us a more precise meaning.

Definition

The *slope* (m) of a line through the two distinct points (x_1, y_1) and (x_2, y_2) is given by the ratio

$$m = \frac{y_2 - y_1}{x_2 - x_1}, \quad x_1 \neq x_2$$

EXAMPLE

Find the slope of a line through the points $(1, -5)$ and $(4, 0)$. If we let $(1, -5)$ be (x_1, y_1) and $(4, 0)$ be (x_2, y_2), then applying the formula we obtain

$$m = \frac{0 + 5}{4 - 1} = \frac{5}{3}$$

The ratio $\frac{y_2 - y_1}{x_2 - x_1}$ is not dependent on the points chosen as long as they are two different points on the line. A given line has only one slope.

The definition of slope, and the fact that the slope of a given line is constant, is the basis for the solution of many problems concerning linear functions.

For instance, if we are given that a line contains the points (x_1, y_1) and (x_2, y_2) and are asked for the equation of the line, we proceed in the following manner.

First we choose any other point on the line and call it (x, y). The slope is now given by the ratio $\frac{y_2 - y_1}{x_2 - x_1}$ or by the ratio $\frac{y - y_1}{x - x_1}$, and since the slope is constant, we can equate these two ratios giving

$$\frac{y - y_1}{x - x_1} = \frac{y_2 - y_1}{x_2 - x_1}$$

which becomes

$$y - y_1 = \left(\frac{y_2 - y_1}{x_2 - x_1}\right)(x - x_1), \quad x_1 \neq x_2$$

This is known as the *two-point form* of the equation of a line.

EXAMPLE

Write the equation in standard form of the line through the points $(3, 5)$ and $(-2, 7)$. Using the two-point form we have

$$y - 5 = \left(\frac{7 - 5}{-2 - 3}\right)(x - 3)$$

Simplifying we obtain

$$y - 5 = -\frac{2}{5}(x - 3)$$

or
$$2x + 5y = 31$$

Check to see that both points are on this line by substituting into the equation.

If we are given the slope m of a line and a point (x_1, y_1) and asked to write the equation, we proceed as follows.

We first choose some other point on the line and call it (x,y). The definition of the slope gives us the ratio $\frac{y - y_1}{x - x_1}$. But since we are given the slope to be m, we can write the equation

$$m = \frac{y - y_1}{x - x_1}$$

or
$$y - y_1 = m(x - x_1)$$

This is the *point-slope form* of the equation of a straight line.

EXAMPLE Write the equation in standard form of the line through the point (2,5) which has a slope of $\frac{2}{5}$. Using the point-slope form gives us

$$y - 5 = \frac{2}{5}(x - 2)$$

which in standard form is

$$2x - 5y = -21$$

Definition The *y-intercept* of a straight line is the ordinate of the point where the line intersects the *y*-axis. In other words, it is the value of y when $x = 0$. We will designate the *y*-intercept as b.

If we are given the slope m of a line and the point $(0,b)$ on the line, then the point-slope form gives us

$$y - b = m(x - 0)$$

or
$$y = mx + b$$

3.4 LINEAR FUNCTIONS

This very useful form of the equation of a straight line is called the *slope-intercept form*.

EXAMPLE If the slope of a line is $-\frac{5}{8}$ and the y-intercept is 6, write the equation in standard form. The slope-intercept form gives us

$$y = -\frac{5}{8}x + 6$$

which in standard form is

$$5x + 8y = 48$$

The slope-intercept form of the equation of a straight line is useful in graphing. We can use the equation from the last example to illustrate.

$$y = -\frac{5}{8}x + 6$$

We know that the y-intercept is 6. Thus the point whose coordinates are (0,6) is on the graph.

The slope is $-\frac{5}{8}$, and since it is a ratio, it does not matter whether we place the negative sign in the numerator or denominator. If we put it in the numerator, we obtain $\frac{-5}{8}$ which indicates that the change in y is -5 while the change in x is 8.

Using these values, we begin at (0,6) and move 8 places in the positive x direction. We then move 5 places in the negative y direction. The point we arrive at is also on the line. Since we now have two points, we may draw the graph of the line.

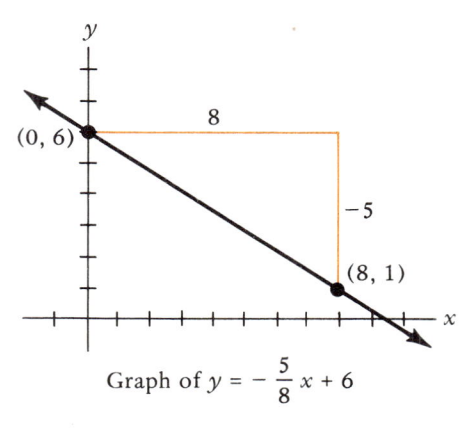

Graph of $y = -\frac{5}{8}x + 6$

EXAMPLE

Find the slope and the *y*-intercept of the line given by the equation $3x - 2y = 9$. Solving for *y* in terms of *x* gives us

$$y = \frac{3}{2}x - \frac{9}{2}$$

so the slope is $m = \frac{3}{2}$ and the *y*-intercept is $b = -\frac{9}{2}$.

If two lines are parallel, how do their slopes compare? Do you see that they would have to be equal?

Exercise 3.4.1

1. Find the equation, in standard form, of the line through each of the following pairs of points:

   a. (2,1) and (−3,9)   b. (−6,2) and (4,−3)

   c. (16,−5) and (4,0)   d. (−8,−1) and (3,−5)

2. Find the equation, in standard form, of the line through each of the given points and having the given slope.

   a. (1,7), $m = 3$ b. (−2,0), $m = \frac{1}{2}$

   c. (0,4), $m = 5$ d. (−5,9), $m = -8$

   e. (3,−1), $m = -\frac{2}{3}$ f. (−2,3), parallel to the *y*-axis

3. Write each of the following equations in slope-intercept form. Specify the slope and *y*-intercept.

   a. $2x + y = 5$ b. $3x + 4y = 1$

   c. $2x - 5y = 9$ d. $3y - 6x = 4$

   e. $5x - 2y = 0$ f. $x - 8y = 1$

4. A line has an *x*-intercept of $x = -2$ and a *y*-intercept of $y = 7$. Find the equation of the line in standard form.

5. Write the equation, in standard form, of the line passing through the given point and parallel to the given line.

   a. (3,1), $y = 2x + 1$ b. (2,0), $3x + y = 5$

c. $(5,-1)$, $2y - 5x = 1$ d. $(3,-10)$, $2x - y = 4$

e. $(-8,1)$, $y = -3$ f. $(4,-9)$, $x = 2$

3.5 SOME ALGEBRA OF FUNCTIONS

Goals

Upon completion of this section you should be able to:

1. Find the sum, difference, quotient, and product functions from given functions.
2. Form the composition of two functions.
3. Find the domain and range of sum, difference, quotient, product, and composite functions.

* *

The set of all functions, together with operations on this set, forms an algebra of functions. In this section we will discuss a few of the basic ideas of this algebra.

Definition

If f and g are functions, then

1. $(f + g)(x) = f(x) + g(x)$ is the *sum* function.
2. $(f - g)(x) = f(x) - g(x)$ is the *difference* function.
3. $\left(\dfrac{f}{g}\right)(x) = \dfrac{f(x)}{g(x)}$ is the *quotient* function.
4. $(f \cdot g)(x) = [f(x)][g(x)]$ is the *product* function.

Take note of the fact that in this definition the domain of the sum, difference, quotient, or product function would contain only those numbers that are in *both* the domain of f *and* in the domain of g. The symbol ∩, which is read as "intersection," is used to describe elements that are common to two sets. If D_f represents the domain of f and D_g represents the domain of g, then $D_f \cap D_g$ would represent the intersection of the two domains. For example, if

$$D_f = \{\text{all real numbers}\}$$

and

$$D_g = \{\text{all positive real numbers}\}$$

then

$$D_f \cap D_g = \{\text{all positive real numbers}\}$$

This intersection is the domain of $f + g$, $f - g$, $\dfrac{f}{g}$, and $f \cdot g$ with the further stipulation that in the quotient function, the denominator g cannot be zero.

EXAMPLE

If $f(x) = 2x - 1$ and $g(x) = 3x + 4$, then find $f + g, f - g, \dfrac{f}{g}$, and $f \cdot g$. Give the domain of each.

$$(f + g)(x) = (2x - 1) + (3x + 4)$$
$$= 5x + 3 \qquad \text{domain: } (-\infty, +\infty)$$

$$(f - g)(x) = (2x - 1) - (3x + 4)$$
$$= -x - 5 \qquad \text{domain: } (-\infty, +\infty)$$

$$\left(\dfrac{f}{g}\right)(x) = \dfrac{2x - 1}{3x + 4} \qquad \text{domain: } \left(-\infty, -\dfrac{4}{3}\right) \cup \left(-\dfrac{4}{3}, +\infty\right)$$

$$(f \cdot g)(x) = (2x - 1)(3x + 4)$$
$$= 6x^2 + 5x - 4 \qquad \text{domain: } (-\infty, +\infty)$$

Exercise 3.5.1

For each of the following pairs of functions find

a. $(f + g)(x)$ b. $(f - g)(x)$ c. $(f \cdot g)(x)$

d. $\left(\dfrac{f}{g}\right)(x)$ e. $\left(\dfrac{g}{f}\right)(x)$

and, in each case, state the domain.

1. $f(x) = x^2$, $g(x) = 3x + 2$
2. $f(x) = x^2 + 2x - 15$, $g(x) = x - 3$
3. $f(x) = \sqrt{x}$, $g(x) = x^2$
4. $f(x) = \sqrt{x + 5}$, $g(x) = \sqrt{x - 5}$
5. $f(x) = \dfrac{1}{x}$, $g(x) = \dfrac{x + 1}{x + 4}$

Definition

If f and g are functions, the composition of f and g or the *composite function* denoted by $(f \circ g)(x)$ is defined as $(f \circ g)(x) = f[g(x)]$.

3.5 SOME ALGEBRA OF FUNCTIONS

The domain of $(f \circ g)$ requires special attention. The definition implies that for $(f \circ g)(x)$ we would substitute x in g and then $g(x)$ in f. This means that $g(x)$ must be in the domain of f, and x must be in the domain of g.

EXAMPLE

If $f(x) = \sqrt{x + 1}$ and $g(x) = 2x - 5$, find $(f \circ g)(x)$ and give its domain.

$$(f \circ g)(x) = f[g(x)]$$
$$= f(2x - 5)$$
$$= \sqrt{(2x - 5) + 1}$$
$$= \sqrt{2x - 4}$$

The domain of f is $[-1, +\infty)$ and the range of g is all reals. Therefore, the domain of $(f \circ g)$ is the set of x such that $g(x) \geq -1$ or $2x - 5 \geq -1$ giving $x \geq 2$. Thus the domain of $(f \circ g)(x)$ is $[2, +\infty)$.

Exercise 3.5.2

Find $(f \circ g)(x)$ and $(g \circ f)(x)$ in each of the following. State the domain in each case.

1. $f(x) = 5x + 1, g(x) = 2x - 3$
2. $f(x) = \dfrac{1}{x}, g(x) = 3x + 1$
3. $f(x) = 2x^2 - x + 1, g(x) = 2x + 1$
4. $f(x) = \sqrt{x + 2}, g(x) = 5x - 2$
5. $f(x) = \sqrt{3x - 1}, g(x) = x^2 + 3$
6. $f(x) = \dfrac{1}{x^2 + 2}, g(x) = \sqrt{2x - 4}$
7. $f(x) = 8, g(x) = 2$
8. $f(x) = x^2, g(x) = \sqrt{x}$
9. $f(x) = 3x + 2, g(x) = \dfrac{x - 2}{3}$
10. $f(x) = x^2 + 5, g(x) = \sqrt{x - 5}$

3.6 INVERSE FUNCTIONS

Goals

Upon completion of this section you should:

1. Know the definition of an inverse relation.
2. Know the definition of an inverse function.
3. Given a function, find its inverse.
4. Impose any necessary restrictions so that a function and its inverse will be inverse functions.

* * * * * * * * * * * * * * * * * * * *

Two ordered pairs of numbers (a,b) and (c,d) are equal if and only if $a = c$ and $b = d$. Therefore, if we take an ordered pair (a,b) and reverse its coordinates, we obtain a new ordered pair (b,a) which is different from the ordered pair (a,b) (unless, of course, $a = b$).

A relation R is, by definition, a set of ordered pairs. If we reverse the coordinates of every ordered pair in this set, we then have a new relation called the *inverse relation* of R.

Definition If the coordinates of each ordered pair (x,y) of the relation R are reversed, the resulting relation is the *inverse of R* and is denoted by R^{-1}.

This definition provides an obvious method for finding the inverse of a relation. Since the order of x and y is interchanged, we will, in the equation representing the relation, simply interchange x and y.

EXAMPLE Find the inverse of $y = x^2$. We interchange x and y obtaining

$$x = y^2$$
or
$$y = \pm\sqrt{x}$$

A table of values will show that the ordered pairs are reversed.

$y = x^2$

x	−3	−2	−1	0	1	2	3
y	9	4	1	0	1	4	9

$y = \pm\sqrt{x}$

x	0	1	4	9
y	0	±1	±2	±3

Note that the domain of $y = x^2$ is the set of all reals and the range is all nonnegative reals; whereas for $y = \pm\sqrt{x}$, the domain is the nonnegative reals and the range is all reals.

EXAMPLE Find the inverse of $y = 2x - 1$. Interchanging x and y gives

$$x = 2y - 1$$

and solving for y we get

$$y = \frac{x + 1}{2}$$

3.6 INVERSE FUNCTIONS

Again a table of values clearly shows ordered pairs reversed.

$y = 2x - 1$

x	0	1	2	3	4
y	−1	1	3	5	7

$y = \dfrac{x+1}{2}$

x	−1	1	3	5	7
y	0	1	2	3	4

In the first example, the relation $y = x^2$ defines a function, while its inverse $y = \pm\sqrt{x}$ does not. However, in the second example the relation $y = 2x - 1$ and its inverse $y = \dfrac{x+1}{2}$ are both functions.

Clearly then, the inverse of a function is not necessarily a function.

Exercise 3.6.1

For each of the following:

a. Find the inverse of the given relation.
b. Is the given relation a function?
c. Is the inverse of the given relation a function?

1. $\{(1,3),(2,5),(-2,3),(0,-1)\}$ 2. $y = x + 1$
3. $y = 5x - 2$ 4. $y = x^2 + 1$
5. $y = 3x^2 - 4$ 6. $y = x^3$
7. $y = (x - 2)^2 + 1$ 8. $3x + 2y = 7$

The idea of an inverse is always related to the *identity element* of a set. For instance, −5 and 5 are additive inverses since their sum equals zero, the additive identity of the set of real numbers. Also, $\dfrac{1}{3}$ and 3 are multiplicative inverses since their product is 1, the multiplicative identity of the set of real numbers. The identity element of an operation on a set is simply that element which leaves all elements of the set unchanged under the operation. Hence the following definitions.

Definition $f(x) = x$ is the *identity element* for the set of functions on the set of real numbers.

Definition The *inverse* of a function $f(x)$ (which we will denote as $f^{-1}(x)$ and read as "f inverse x") is a function such that
$$(f \circ f^{-1})(x) = x \quad \text{and} \quad (f^{-1} \circ f)(x) = x$$
for all x in the domain of $f(x)$.

Note first that this last definition interchanges x and y in all ordered pairs (x,y) of $f(x)$. If (a,b) is an ordered pair of f, then $f(a) = b$ and $(f \circ f^{-1})(b) = f[f^{-1}(b)] = b$ imply that $f^{-1}(b) = a$, making (b,a) an ordered pair of $f^{-1}(x)$.

However, this definition does more than reverse the roles of x and y. It also requires that $f(x)$ and $f^{-1}(x)$ both be functions and that the range of $f(x)$ be the domain of $f^{-1}(x)$. Hence, we must be cautioned that the "inverse of a function" is not necessarily "an inverse function."

In general, any relation can be used to define a function if the domain or range (or both) is restricted. It is clear from the definition of functions that these imposed restrictions are designed so that the relation will yield one and only one image of x for each x in the domain.

Carefully consider the following example to reinforce the finer points of the definitions and discussion.

EXAMPLE Given the function $y = f(x) = x^2$:
a. Find the inverse of $y = x^2$.
b. Is the inverse of $y = x^2$ a function? If not, restrict the range of the inverse of $y = x^2$ so that it is a function.
c. Determine if $y = x^2$ and its inverse (with restricted range) are inverse functions. If not, impose the necessary restrictions so that they are inverse functions.

a. We interchange x and y, then solve for y to get $y = \pm\sqrt{x}$ as the inverse of $y = x^2$.

b. $y = \pm\sqrt{x}$ is not a function, but $y = \pm\sqrt{x}$ such that $y \geq 0$ is a function. Hence $y = \sqrt{x}$ is a function.

c. Now we must determine if $y = x^2$ and $y = \sqrt{x}$ are inverse functions. For the sake of notation we will use $f(x) = x^2$ and $g(x) = \sqrt{x}$. Our question then becomes "Is $g(x) = f^{-1}(x)$?" We find

3.6 INVERSE FUNCTIONS

and
$$(f \circ g)(x) = f[g(x)]$$
$$= f(\sqrt{x})$$
$$= (\sqrt{x})^2$$
$$= x$$
$$(g \circ f)(x) = g[f(x)]$$
$$= g(x^2)$$
$$= \sqrt{x^2}$$
$$= |x|$$

Thus f and g are *not* inverse functions.

Since we want $(g \circ f)(x) = x$ and have instead $(g \circ f)(x) = |x|$, we note that $|x| = x$ if and only if $x \geqslant 0$. Using the restriction $x \geqslant 0$ on the domain of $f(x)$, we now can state that the functions $f(x) = x^2$ (such that $x \geqslant 0$) and $f^{-1}(x) = \sqrt{x}$ are inverse functions.

A graphical representation of $f(x) = x^2$, $x \geqslant 0$, and $f^{-1}(x) = \sqrt{x}$ follows. $y = x$ is on the same coordinate axis.

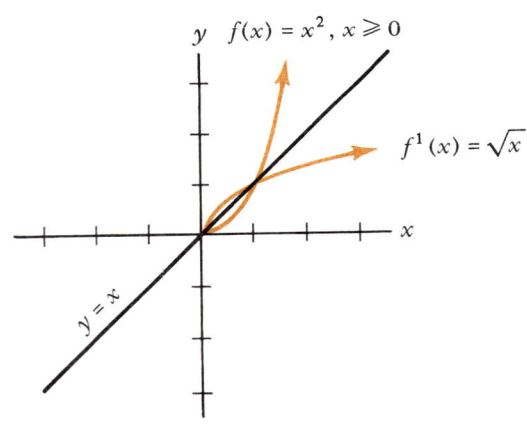

Notice that the given function and its inverse are reflections in the line $y = x$.

Exercise 3.6.2

For the following:

a. Find the inverse of the given function.

b. Impose the necessary restrictions so that the given function and its inverse are inverse functions.

1. $f(x) = 2x - 7$
2. $f(x) = \frac{1}{2}x + \frac{2}{3}$
3. $f(x) = \frac{1}{x}, \; x \neq 0$
4. $f(x) = \frac{x + 3}{x}, \; x \neq 0$
5. $f(x) = x^2 - 3$
6. $f(x) = \sqrt{x^2 + 4}$
7. $f(x) = x^3$
8. $f(x) = \frac{1}{x^2}, \; x \neq 0$

9. In the last example discussed in this preceding section, could we have used $y = x^2$ and $y = -\sqrt{x}$ to obtain inverse functions? Explain.

CHAPTER REVIEW

Give the domain and range of each of the following relations. In each case state if the relation is a function.

1. $y = 2x^2$
2. $y = \frac{1}{3x}$
3. $x = y^2 + 1$
4. $x = \frac{3}{y}$
5. $y = \sqrt{4 - x^2}$
6. $y = \frac{x - 1}{x + 5}$
7. $y = \frac{1}{\sqrt{x^2 - 1}}$
8. $y = |x^2 - 1|$

9. Find the distance between the points $(-9, 3)$ and $(5, -7)$.
10. If the line segment AB has midpoint $(5, -2)$ and the coordinates of B are $(-8, 4)$, find the coordinates of A.
11. Put the equation of the circle $x^2 + y^2 - 6x + 2y = 15$ in standard form. Give the center and radius.

For each of the relations given in Problems 12 to 20, give the domain and range. State whether or not the relation is a function, and graph it.

12. $y = 5x - 2$
13. $f(x) = x^2 - 2x + 3$
14. $x^2 + y^2 - 4x = 0$
15. $x = y^2 + 6y + 5$
16. $x = 4$
17. $y = |x - 2| + 3$
18. $f(x) = x^3 - 4x$
19. $x = |y + 1|$
20. $f(x) = \sqrt{x^2 - 1}$

In Problems 21 to 24, put the equation in standard form and classify the curve. Sketch the graph.

21. $25x^2 + 9y^2 + 150x - 90y + 225 = 0$
22. $4x^2 + 4y^2 - 4x - 48y + 45 = 0$
23. $25y^2 - 16x^2 + 50y + 64x - 439 = 0$
24. $y^2 - 8y - 2x + 6 = 0$

25. Classify the curve represented by the equation $x^2 + y^2 + 6x - 8y + 25 = 0$.

26. Find the equation, in standard form, of the line through the points $(4,-6)$ and $(-3,5)$.

27. Find the equation, in standard form, of the line through the point $(2, -9)$ and having a slope of $-\dfrac{3}{5}$.

28. Write the equation $2x - 5y = 10$ in slope-intercept form.

29. Find the equation of a line, in standard form, whose slope is 5 and has a y-intercept of -3.

30. Write the equation, in slope-intercept form, of the line passing through the point $(-2,7)$ which is parallel to the line $3x + 8y = 1$.

Given $f(x) = 3x^2 + 2x$ and $g(x) = x - 1$, in Problems 31 to 35 find the indicated function and give its domain.

31. $(f + g)(x)$
32. $(f - g)(x)$
33. $(f \cdot g)(x)$
34. $\left(\dfrac{f}{g}\right)(x)$
35. $(f \circ g)(x)$

Given $f(x) = \dfrac{1}{x + 1}$ and $g(x) = \dfrac{2}{x}$, in Problems 36 to 40 find the indicated function and give its domain.

36. $(f + g)(x)$
37. $(f - g)(x)$
38. $(f \cdot g)(x)$
39. $\left(\dfrac{f}{g}\right)(x)$
40. $(f \circ g)(x)$

Using the definition of inverse functions, prove that the following pairs of functions are inverses of each other:

41. $f(x) = 7x + 3,\ g(x) = \dfrac{x - 3}{7}$

42. $f(x) = \sqrt{5x + 1},\ x \geq -\dfrac{1}{5};\ g(x) = \dfrac{x^2 - 1}{5},\ x \geq 0$

Find the inverse function of each of the following functions:

43. $f(x) = 3x^2 - 4,\ x \leq 0$
44. $f(x) = \sqrt{2x + 3},\ x \geq -\dfrac{3}{2}$
45. $f(x) = \dfrac{1}{x + 3},\ x > 0$

PRACTICE TEST

1. Give the domain and range of each of the following relations. In each case state if the relation is a function.

   a. $y = \sqrt{2x - 5}$
   b. $x^2 + y^2 = 9$
   c. $y = |x - 5|$

2. Given the points $A:(2,8)$ and $B:(-7,3)$

   a. Find the distance between A and B.
   b. Find the midpoint of $\overline{AB}$.
   c. Find the equation, in standard form, of the line which passes through points A and B.

3. Put the equation of the circle $x^2 + y^2 - 10x + 4y = 7$ in standard form. Give the center and radius.

4. Graph the following relations. Give the domain and range of each and state whether or not the relation is a function.

   a. $y = x^2 - 4$
   b. $y = \sqrt{x + 2}$

5. Classify and sketch: $16x^2 - 4y^2 - 64x - 32y - 64 = 0$.

6. Write the equation $8x - 3y = 12$ in slope-intercept form.

7. Given $f(x) = \dfrac{1}{2x - 3}$ and $g(x) = 3x - 1$, find the indicated function and give its domain.

   a. $(f + g)(x)$
   b. $\left(\dfrac{f}{g}\right)(x)$
   c. $(f \circ g)(x)$

8. Find the inverse function of each of the following functions:

   a. $f(x) = 6x - 5$
   b. $f(x) = \sqrt{x + 4}$, $x \geq -4$

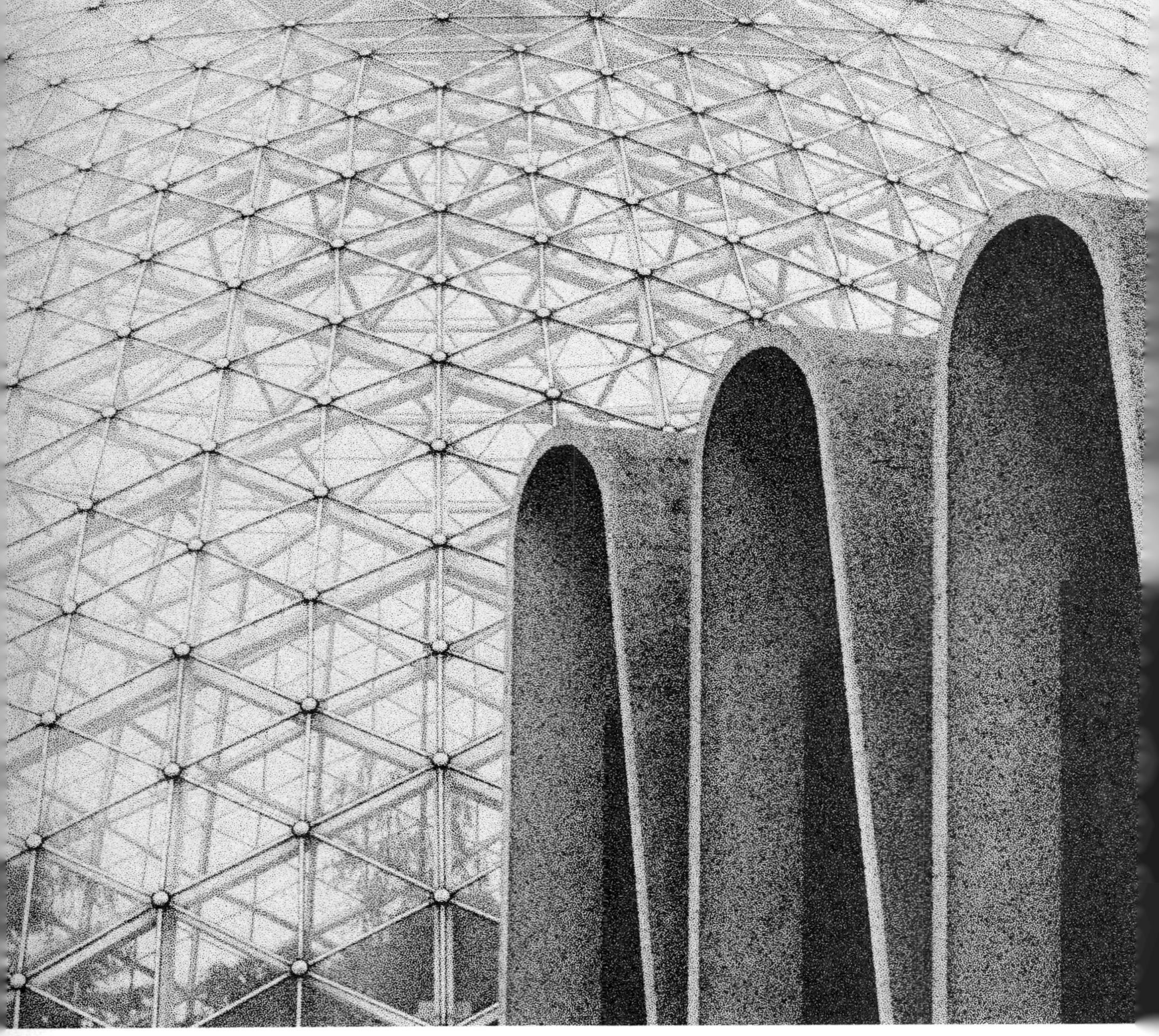

Systems of Equations and Inequalities

4

4

In preceding chapters we have discussed the coordinate system and graphing, solving equations in one variable, solving inequalities in one variable, etc. We now expand these ideas to include sets of equations in two or three variables and look for common solutions to the set. These ideas become useful tools in mathematics as well as in other areas where more than one variable is involved in an equation.

4.1 SOLVING SYSTEMS OF EQUATIONS IN TWO VARIABLES BY GRAPHING

Goals

Upon completion of this section you should be able to:

1. Sketch the graphs of two equations on the same coordinate plane.
2. Determine the common solution to a system of equations from this graph.

* *

A system of equations in two variables is a set of two or more equations. We know that an equation such as $x + y = 5$ has infinitely many solutions such as $(3,2),(6,-1)$. The equation $2x + y = 8$ also has infinitely many solutions such as $(0,8),(4,0)$.

The solution to the system

$$\begin{cases} x + y = 5 \\ 2x + y = 8 \end{cases}$$

is the set of all ordered pairs that are solutions to both equations simultaneously. In other words, is there an ordered pair that makes the statements $x + y = 5$ and $2x + y = 8$ both true?

If we graph both equations on the same coordinate plane, the point or points of intersection will be the solution of the system. These graphs must be extremely accurate in order to arrive at the solution set.

EXAMPLE Solve by graphing

$$\begin{cases} x + y = 5 \\ 2x + y = 8 \end{cases}$$

Set up a table of values for each equation and sketch their graphs. (We should recognize each of these equations as graphically representing a straight line.)

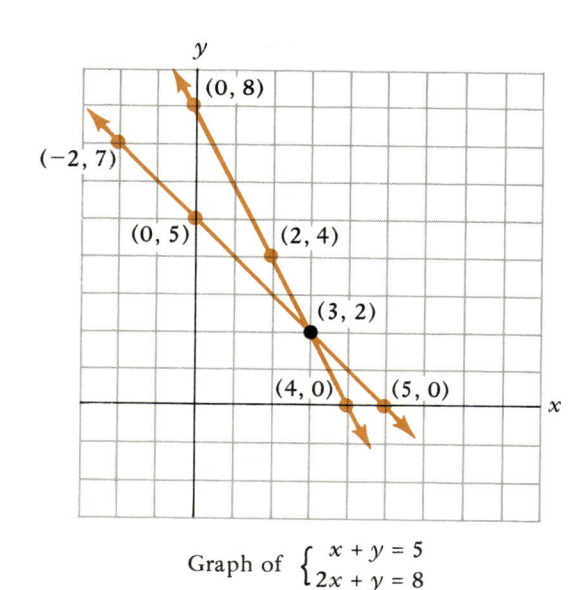

$x + y = 5$

x	-2	0	5
y	7	5	0

$2x + y = 8$

x	0	2	4
y	8	4	0

Graph of $\begin{cases} x + y = 5 \\ 2x + y = 8 \end{cases}$

The point of intersection is (3,2) and should be labeled as in the figure shown. We can check the correctness of our solution by substituting (3,2) into both of the equations to see that it is actually a solution.

4.1 SOLVING EQUATIONS IN TWO VARIABLES BY GRAPHING

EXAMPLE Solve by graphing

$$\begin{cases} x - y = -1 \\ y = x^2 - 2x + 1 \end{cases}$$

These equations represent a straight line and a parabola, respectively. Again set up tables of values and graph each equation.

$x - y = -1$

x	-3	1	5
y	-2	2	6

$y = x^2 - 2x + 1$

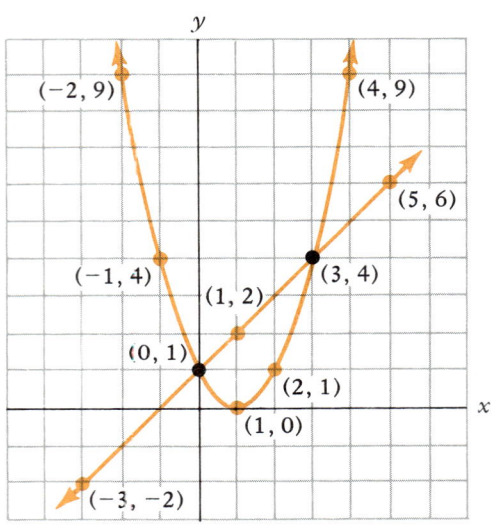

x	-2	-1	0	1	2	3	4
y	9	4	1	0	1	4	9

Graph of $\begin{cases} x - y = -1 \\ y = x^2 - 2x + 1 \end{cases}$

We see that the solutions are (0,1) and (3,4). These check in both equations.

Exercise 4.1.1

Solve the following systems by graphing:

1. $\begin{cases} x + y = 2 \\ 2x - y = 1 \end{cases}$
2. $\begin{cases} 3x + y = 0 \\ 2x - y = -5 \end{cases}$
3. $\begin{cases} 3x + y = -9 \\ x - 2y = 4 \end{cases}$
4. $\begin{cases} 3x + y = 8 \\ 2x - y = 7 \end{cases}$
5. $\begin{cases} 3x + 2y = 7 \\ 2x - 3y = -4 \end{cases}$
6. $\begin{cases} x - y = -2 \\ y = x^2 \end{cases}$
7. $\begin{cases} x + y = -1 \\ y = x^2 + x - 1 \end{cases}$
8. $\begin{cases} y = x^2 - 2x + 3 \\ x + y = 5 \end{cases}$

9. Given a system of two linear equations, discuss the possibilities for the solution set.

10. Given a system of one straight line and one parabola, discuss the possibilities for the solution set.

4.2 THE ALGEBRAIC SOLUTION OF A SYSTEM OF TWO LINEAR EQUATIONS

Goals

Upon completion of this section you should be able to:

1. Solve a system of two linear equations by the substitution method.
2. Solve a system of two linear equations by the addition method.
3. Determine if a system of two linear equations is independent, inconsistent, or dependent.

* *

We will discuss two algebraic methods of solving a system of two first-degree equations in two variables.

The first method to be discussed is the method of *substitution*. This method involves solving for one unknown (variable) in terms of the other in one of the two equations and then substituting this expression into the other equation.

EXAMPLE   Solve by the substitution method

$$\begin{cases} 2x + 3y = 1 \\ x - 2y = 4 \end{cases}$$

Step 1 We must solve for one of the variables in one of the equations. We can choose either x or y in either the first or second equation. Our choice can be based on obtaining the simplest expression. In this case we will solve for x in the second equation obtaining

$$x = 4 + 2y$$

(Notice that any other choice would have resulted in a fraction being involved.)

Step 2 Substitute this value of x into the other equation. In this case the other equation is the first equation

$$2x + 3y = 1$$

Substituting $(4 + 2y)$ for x we obtain

$$2(4 + 2y) + 3y = 1$$

(Note that this equation has only one variable.)

Step 3 Solve this equation for the variable y.

$$2(4 + 2y) + 3y = 1$$
$$8 + 4y + 3y = 1$$
$$8 + 7y = 1$$
$$7y = -7$$
$$y = -1$$

Step 4 Substitute $y = -1$ into either equation to find the corresponding value for x. Since we have already solved the second equation for x in terms of y, we may use it.

$$\begin{aligned} x &= 4 + 2y \\ &= 4 + 2(-1) \\ &= 4 - 2 \\ &= 2 \end{aligned}$$

Thus we have the solution $(2, -1)$.

Step 5 Check the solution in both equations. Remember that the solution for a system must be a solution of each equation in the system. Since

$$2x + 3y = 2(2) + 3(-1) = 4 - 3 = 1$$

and $\quad x - 2y = 2 - 2(-1) = 2 + 2 = 4$

we see that the solution $(2,-1)$ does check.

EXAMPLE   Solve by substitution

$$\begin{cases} 2x + 3y = 7 \\ 4x + 3y = 8 \end{cases}$$

Step 1 We will obtain a fractional expression in any case, so one choice is as easy as another. We will solve for x in the first equation.

$$x = \frac{7 - 3y}{2}$$

Step 2 Substitute the expression $\left(\dfrac{7 - 3y}{2}\right)$ for x in the second equation.

$$4x + 3y = 8$$

$$4\left(\frac{7 - 3y}{2}\right) + 3y = 8$$

Step 3 Solve the equation.

$$4\left(\frac{7 - 3y}{2}\right) + 3y = 8$$

$$2(7 - 3y) + 3y = 8$$

$$14 - 6y + 3y = 8$$

$$-3y = -6$$

$$y = 2$$

Step 4 Substitute $y = 2$ in either of the equations. If we use the first equation, we obtain

$$2x + 3y = 7$$

$$2x + 3(2) = 7$$

4.2 THE ALGEBRAIC SOLUTION OF TWO LINEAR EQUATIONS

$$2x + 6 = 7$$
$$2x = 1$$
$$x = \frac{1}{2}$$

Step 5 Checking we find that the ordered pair $\left(\frac{1}{2}, 2\right)$ satisfies both equations and is thus the solution to the system.

Exercise 4.2.1

Solve by the substitution method.

1. $\begin{cases} 2x + y = 4 \\ 3x - 2y = -1 \end{cases}$
2. $\begin{cases} x + 5y = 2 \\ 2x + 3y = -3 \end{cases}$
3. $\begin{cases} x + y = 5 \\ 2x - y = -5 \end{cases}$

4. $\begin{cases} 2x + 3y = 1 \\ 4x + 5y = 3 \end{cases}$
5. $\begin{cases} 5x - y = 11 \\ 2x - 3y = 7 \end{cases}$
6. $\begin{cases} x + 3y = -1 \\ 2x + 5y = 1 \end{cases}$

7. $\begin{cases} 3x - 2y = -6 \\ x + 2y = 22 \end{cases}$
8. $\begin{cases} 2x + 3y = 2 \\ 3x - y = -19 \end{cases}$

The next method we will discuss is the *addition method*. It is based on two facts we have used previously.

First, we know that the solutions to an equation are not changed if every term of that equation is multiplied by a nonzero number, and secondly, we know that if we add the same or equal quantities to both sides of an equation, the results are still equal.

EXAMPLE

Solve by addition
$$\begin{cases} 2x + y = 5 \\ 3x + 2y = 6 \end{cases}$$

Step 1 Our purpose is to add the two equations and eliminate one of the variables so that we can solve the resulting equation in one variable. If we add the equations as they are, we will not eliminate a variable. This means we must first multiply each side of one or both of the equations by a number or numbers that will lead to the elimination of one of the variables when the equations are added.

If we decide to eliminate x, we can multiply each side of the first equation by 3 and each side of the second equation by -2 giving

$$\begin{cases} 6x + 3y = 15 \\ -6x - 4y = -12 \end{cases}$$

Step 2 Add the equations.

$$\begin{cases} 6x + 3y = 15 \\ -6x - 4y = -12 \end{cases}$$
$$-y = 3$$

Step 3 Solve the resulting equation for the variable y.

$$-y = 3$$
$$y = -3$$

Step 4 Find the other variable by substituting this value into one of the original equations. We will choose the first equation.

$$2x + y = 5$$
$$2x + (-3) = 5$$
$$2x - 3 = 5$$
$$2x = 8$$
$$x = 4$$

Step 5 If we check the ordered pair $(4,-3)$ in both equations, we see that it is a solution of the system.

Notice that this method also gives us a choice as to how we may proceed. In the example just given, we could have decided to eliminate y. Then in Step 1 we would have multiplied each side of the first equation by -2 and left the second equation as it is.

$$\begin{cases} -4x - 2y = -10 \\ 3x + 2y = 6 \end{cases}$$

We may choose our course of action on the basis of obtaining our solution by the easiest method, but either method will yield the same solution.

Exercise 4.2.2

Solve the following by the addition method:

1. $\begin{cases} 2x + y = 1 \\ x - y = 5 \end{cases}$
2. $\begin{cases} 2x - 3y = 5 \\ 2x + y = 9 \end{cases}$
3. $\begin{cases} 2x + y = 1 \\ 3x - 2y = 5 \end{cases}$
4. $\begin{cases} 2x + 3y = 21 \\ 3x + 2y = 19 \end{cases}$
5. $\begin{cases} 5x - 3y = 10 \\ 2x + y = 4 \end{cases}$
6. $\begin{cases} 2x - y = 13 \\ 3x + 5y = 13 \end{cases}$
7. $\begin{cases} 5x + 2y = 2 \\ 2x + 3y = 14 \end{cases}$
8. $\begin{cases} 2x + 5y = -6 \\ 3x + 4y = 5 \end{cases}$

A system of two linear equations does not always have a unique solution, and we should be aware of the other possibilities. (See Problem 9 in Exercise 4.1.1.)

Since we are dealing with linear equations, we may think of the lines represented by the equations. Two linear equations will represent one of three possible situations.

1. *Independent equations:* The two lines intersect in a point. In this case we have a unique solution.
2. *Inconsistent equations:* The two lines are parallel. In this case we have no solution.
3. *Dependent equations:* The two equations give the same line. In this case any solution of one equation is a solution of the other.

We can recognize any of the three situations by proceeding to work the problem either by the substitution or addition method.

1. If the equations are *independent*, we will obtain a *unique solution*.
2. If the equations are *inconsistent*, we will obtain a *contradiction*.
3. If the equations are *dependent*, we will obtain an *identity*.

EXAMPLE Solve the system

$$\begin{cases} x + y = 6 \\ x + y = 8 \end{cases}$$

If we multiply both sides of the second equation by −1 and add, we obtain

$$0 = -2$$

Since this is a *contradiction*, the equations are *inconsistent* and their graphs would be parallel lines. This system has no solution.

EXAMPLE Solve the system

$$\begin{cases} x + y = 6 \\ 2x + 2y = 12 \end{cases}$$

Multiplying both sides of the first equation by −2 and adding yields

$$0 = 0$$

Since this is an *identity*, the equations are *dependent* and represent the same line on the coordinate plane. Therefore, any ordered pair that satisfies one equation will also satisfy the other. There are infinitely many solutions. It would therefore be impossible to list all the solutions; however, we would like to have some way of indicating what these solutions are. A convenient tool which is used for this purpose is *set notation*.

Set notation is a precise shorthand technique of stating in a concise way what might otherwise take several written statements to explain. The solution to this example may be written as

$$\{(x,y) \mid x + y = 6\}$$

and at first glance might seem quite complex. Remember, however, that the symbols used are a shorthand technique and therefore need translation. The braces { } mean "the set," and the verticle bar | means "such that." Thus the solution would be read as "the set of all ordered pairs (x,y) such that $x + y = 6$."

EXAMPLE We often use systems of equations to solve verbal problems.

It took a boat 3 hours to travel 75 miles downstream and 5 hours to make the return trip upstream. Find the average speed of the boat and stream.

4.2 THE ALGEBRAIC SOLUTION OF TWO LINEAR EQUATIONS

We must use a formula from physics to solve this problem. The formula states

$$\text{rate} \times \text{time} = \text{distance}$$

If we first consider the trip downstream, we recognize that the rate of travel will be the sum of the speed of the boat and the speed of the stream. If we let

$$b = \text{speed of the boat}$$

and

$$s = \text{speed of the stream}$$

then

$$(b + s)(3) = 75$$

The rate of travel for the trip upstream will be the speed of the boat decreased by the speed of the stream. Thus

$$(b - s)(5) = 75$$

We now have the system

$$\begin{cases} 3b + 3s = 75 \\ 5b - 5s = 75 \end{cases}$$

Solving the system yields $b = 20$ mph and $s = 5$ mph.

Exercise 4.2.3

Classify each of the following systems as independent, inconsistent, or dependent. If the system is independent, find its solution.

1. $\begin{cases} 2x + y = 1 \\ 3x - y = 9 \end{cases}$

2. $\begin{cases} x + y = 4 \\ x - y = 6 \end{cases}$

3. $\begin{cases} x + 2y = 9 \\ 3x - y = -1 \end{cases}$

4. $\begin{cases} 2x - y = 1 \\ 6x - 3y = 3 \end{cases}$

5. $\begin{cases} 3x - y = 2 \\ 2x + 3y = -6 \end{cases}$

6. $\begin{cases} 3x - 2y = 3 \\ 6x - 4y = 1 \end{cases}$

7. $\begin{cases} 6x - 4y = 2 \\ 3x - 2y = 1 \end{cases}$

8. $\begin{cases} x - 2y = 15 \\ 3x + y = 3 \end{cases}$

9. $\begin{cases} 3x + 2y = 5 \\ 2x + y = 1 \end{cases}$

10. $\begin{cases} 2x + y = 5 \\ 10x + 5y = 10 \end{cases}$

11. $\begin{cases} 2x + 3y = 10 \\ 6x - 2y = -3 \end{cases}$

12. $\begin{cases} x + 3y = 1 \\ 2x - 9y = -8 \end{cases}$

13. $\begin{cases} 2x - 4y = 6 \\ 3x - 6y = 9 \end{cases}$ 14. $\begin{cases} 6x + 3y = 3 \\ 10x + 5y = 15 \end{cases}$ 15. $\begin{cases} 5x + 2y = -1 \\ 4x + 3y = -12 \end{cases}$

16. The sum of two numbers is 147 and their difference is 19. Find the numbers.

17. An airliner took 6 hours to travel 2,400 miles from New York to Las Vegas. The return trip took 4 hours. Find the speed of the plane and the speed of the wind.

18. A professor has 85 students enrolled in two classes; 49 of these students are freshmen. If two-thirds of the first class and one-half of the second class are freshmen, how many students are in each class?

19. A boat travels 15 miles per hour downstream and 9 miles per hour upstream. Find the speed of the current and the speed of the boat.

20. How much of a 40% solution of alcohol and how much of an 80% solution should be mixed to give 40 liters of a 50% solution?

4.3 THE ALGEBRAIC SOLUTION OF THREE EQUATIONS WITH THREE VARIABLES

Goals

Upon completion of this section you should be able to:

1. Solve a system of three first-degree equations in three variables by addition.

* *

A first-degree polynomial equation in three variables is the equation of a plane in three-dimensional space. Three such planes in space might intersect in a single point. Of course, there are other possibilities (such as all three parallel, etc.), and you may wish to list them. In this section we are interested only in those that intersect in a single point (that is, those that have a unique solution) and in an algebraic method of finding the solution.

A point in three space is represented by an *ordered triple* of numbers. We use x, y, and z as the three variables: the ordered triple is (x,y,z). The ordered triple $(2,3,-1)$ represents a point such that $x = 2$, $y = 3$, and $z = -1$.

The method we will use reduces a system of three equations in three variables to two equations in two variables by addition. This method is best illustrated by example.

EXAMPLE   Solve the system

$$\begin{cases} 2x + 3y - z = 11 & (1) \\ x + 2y + z = 12 & (2) \\ 3x - y + 2z = 5 & (3) \end{cases}$$

Step 1   Choose any two equations and eliminate any one of the three variables by addition.

Note that this gives a wide range of choices. We can choose on the basis of the variable that is easiest to eliminate. In this case we will choose to eliminate z by adding equations (1) and (2). Equation (1) added to equation (2) yields

$$3x + 5y = 23$$

Step 2   Choose two *other* equations and eliminate the *same* variable by addition.

The choices here are not as broad because we must eliminate the same variables as in Step 1, but cannot use the same two equations as before. In this example we must eliminate z and can use equations (1) and (3) or equations (2) and (3). We will choose equations (1) and (3) and eliminate z by multiplying both sides of equation (1) by 2 and adding the result to equation (3).

Twice equation (1) added to equation (3) yields

$$7x + 5y = 27$$

Step 3   Solve the system of two equations in two variables that results from Step 1 and Step 2.

$$\begin{cases} 3x + 5y = 23 \\ 7x + 5y = 27 \end{cases}$$

Using either substitution or addition to solve this system, we find

$$x = 1 \quad \text{and} \quad y = 4$$

Step 4   Use the values from Step 3 in any one of the original equations to find the other variable.

Using equation (2) and $x = 1$ and $y = 4$, we have

SYSTEMS OF EQUATIONS AND INEQUALITIES

$$x + 2y + z = 12$$
$$1 + 2(4) + z = 12$$
$$z = 3$$

Step 5 Check the solution in all three of the original equations to see that each equation is satisfied.

In our example, the solution (1,4,3) checks in all three equations.

EXAMPLE Solve the system

$$\begin{cases} x - y + 2z = 6 & (1) \\ 2x + 3y - z = -3 & (2) \\ 3x + 2y + 2z = 5 & (3) \end{cases}$$

Step 1 Eliminate y by multiplying both sides of equation (1) by 3 and adding the result to equation (2). Three times equation (1) added to equation (2) gives

$$5x + 5z = 15$$

Step 2 Eliminate y by multiplying both sides of equation (1) by 2 and adding the result to equation (3). This yields

$$5x + 6z = 17$$

Step 3 Solve the system

$$\begin{cases} 5x + 5z = 15 \\ 5x + 6z = 17 \end{cases}$$

We find the solution to be

$$x = 1 \quad \text{and} \quad z = 2$$

Step 4 Substitute $x = 1$ and $z = 2$ in equation (1) to solve for y. We find the solution to be

$$y = -1$$

Step 5 The ordered triple (1,−1,2) checks in all three equations.

If one of the variables is missing from one or more of the equations, finding the solution becomes easier.

EXAMPLE

Solve the system

$$\begin{cases} x + z = 3 & (1) \\ 2x + y + z = 3 & (2) \\ 3x - y + 2z = 8 & (3) \end{cases}$$

Note that equation (1) contains only two of the variables. We can eliminate y using equations (2) and (3), and then we will have two equations in two variables (x and z). Equation (2) added to equation (3) gives

$$5x + 3z = 11$$

Now we solve the system

$$\begin{cases} x + z = 3 \\ 5x + 3z = 11 \end{cases}$$

We obtain as our solution

$$x = 1 \quad \text{and} \quad z = 2$$

Substituting these values into equation (2) yields

$$y = -1$$

The solution $(1, -1, 2)$ checks in all three equations.

EXAMPLE

The equation of a vertical parabola can be put in the form $y = ax^2 + bx + c$. Find the equation of the vertical parabola that passes through the points $(-3, 10)$, $(-2, 3)$, and $(1, 6)$.

We must find the values of a, b, and c. Since each point must satisfy the equation, we have

for $(-3, 10)$: $\quad 10 = a(-3)^2 + b(-3) + c$

for $(-2, 3)$: $\quad 3 = a(-2)^2 + b(-2) + c$

and for $(1, 6)$: $\quad 6 = a(1)^2 + b(1) + c$

These simplify to the system

$$\begin{cases} 9a - 3b + c = 10 \\ 4a - 2b + c = 3 \\ a + b + c = 6 \end{cases}$$

The solution yields $a = 2$, $b = 3$, $c = 1$. Thus the desired equation is $y = 2x^2 + 3x + 1$.

EXAMPLE Three technicians can together complete a job in 3 hours. Technicians A and B working together can finish in 5 hours. If technicians A and C work together, they can complete it in 4½ hours. How long would it take each one working alone to complete the job?

We let a represent the rate at which technician A works (i.e., the portion of the job completed in 1 hour). The same is done for the others.

$$a = \text{rate for technician } A$$
$$b = \text{rate for technician } B$$
$$c = \text{rate for technician } C$$

Since it takes all three technicians 3 hours to complete the job, then in 1 hour they must complete one-third of the job. Thus

$$a + b + c = \frac{1}{3}$$

Also
$$a + b = \frac{1}{5}$$

and
$$a + c = \frac{2}{9}$$

If we clear fractions, we have the system

$$\begin{cases} 3a + 3b + 3c = 1 \\ 5a + 5b = 1 \\ 9a + 9c = 2 \end{cases}$$

Solving, we obtain $a = \dfrac{4}{45}$, $b = \dfrac{1}{9}$, $c = \dfrac{2}{15}$.

Therefore technician A can complete the job in 11¼ hours, technician B in 9 hours, and technician C in 7½ hours.

4.3 THE ALGEBRAIC SOLUTION OF THREE EQUATIONS

Exercise 4.3.1

Solve the following systems:

1. $\begin{cases} x + y + z = 6 \\ 2x - y + z = 3 \\ x - y + 2z = 5 \end{cases}$
2. $\begin{cases} x + 2y + z = 0 \\ x - 3y - z = -2 \\ x + y - z = -2 \end{cases}$
3. $\begin{cases} 2x + y + z = 0 \\ 3x - 2y - z = -11 \\ x - y + 2z = 3 \end{cases}$
4. $\begin{cases} x - y + z = 8 \\ 5x + 4y - z = 7 \\ 2x + y - 3z = -7 \end{cases}$
5. $\begin{cases} x + 5y - 2z = 13 \\ 6x + y + 3z = 4 \\ x - y + 2z = -5 \end{cases}$
6. $\begin{cases} x + y = 6 \\ 2x - y + z = 7 \\ x + y - 3z = 12 \end{cases}$
7. $\begin{cases} 3x + 4y + z = -2 \\ y + z = 1 \\ 2x - y - z = -5 \end{cases}$
8. $\begin{cases} y - z = -3 \\ x + y = 1 \\ 2x + 3y + z = 1 \end{cases}$

9. A child has $4.50 in nickles, dimes, and quarters. He has a total of 28 coins and the number of dimes is twice the number of nickels. How many of each coin does the child have?

10. The equation of a circle can be put in the form $x^2 + y^2 + ax + by + c = 0$. If the circle contains the points $(-1,0)$, $(0,1)$, and $(1,3)$, find the values of a, b, and c and write the equation of the circle.

11. A chemist wishes to make 9 liters of a 30% acid solution by mixing three solutions of 5%, 20%, and 50%. How much of each must she use if she uses twice as much 50% solution as she does 5% solution?

12. A swimming pool can be filled by two pipes A and B and drained by a third pipe C. If the drain is closed and both pipes A and B are open, it takes 6 hours to fill the pool. If A and B are open and the drain is also open, it takes 10 hours to fill the pool. If only pipes A and C are open, it takes 15 hours to fill the pool. How long would it take to drain the pool if A and B were closed and C were open?

4.4 MATRICES

Goals

Upon completion of this section you should be able to:

1. Define an m by n matrix.
2. Use the elementary row operations to change a matrix to triangular form.

3. Use the triangular form of an augmented matrix to solve a system of first-degree equations.

* *

A *matrix* is any array of numbers called elements arranged in horizontal rows and vertical columns. The topic of matrices is very important in many branches of mathematics and a general matrix theory can be developed which is an "algebra of matrices" having operations and relations. However, in this section we will introduce only those basic ideas that are useful in helping to solve systems of equations.

Definition If m and n are positive integers, an *m by n matrix* is an array of the form

$$\begin{bmatrix} a_{11} & a_{12} & a_{13} & \cdots & a_{1n} \\ a_{21} & a_{22} & a_{23} & \cdots & a_{2n} \\ a_{31} & a_{32} & a_{33} & \cdots & a_{3n} \\ \vdots & & & & \vdots \\ a_{m1} & a_{m2} & a_{m3} & \cdots & a_{mn} \end{bmatrix}$$

where a_{ij} represents an element in row i and column j.

With any system of linear equations, there are associated two matrices; first the *matrix of coefficients*, and second, the *augmented matrix*.

For example, given the system

$$\begin{cases} 2x + 3y - z = 11 \\ x + 2y + z = 12 \\ 3x - y + 2z = 5 \end{cases}$$

the matrix of coefficients is

$$\begin{bmatrix} 2 & 3 & -1 \\ 1 & 2 & 1 \\ 3 & -1 & 2 \end{bmatrix}$$

4.4 MATRICES

Notice that the first row corresponds to the coefficients of the first equation, the second row to the second equation, and the third row to the third equation. Also, the first column contains the x coefficients, the second column the y coefficients, and the third column the z coefficients. The augmented matrix is

$$\begin{bmatrix} 2 & 3 & -1 & 11 \\ 1 & 2 & 1 & 12 \\ 3 & -1 & 2 & 5 \end{bmatrix}$$

This matrix is the same as the first with the exception that it contains a fourth column which is composed of the constant terms in the equation.

We will work in this section with the augmented matrix.

Any two systems of equations that have the same solution set are said to be *equivalent*. It can be shown that certain operations on the augmented matrix will result in a matrix of an equivalent system and therefore not change the solution of the system. The operations we will use are referred to as *elementary row operations*.

Theorem

The following operations on an augmented matrix will result in a matrix of an equivalent system:

1. **Interchanging any two rows.**
2. **Multiplying all elements of a row by a constant k ($k \neq 0$).**
3. **Multiplying the elements of a row by a constant and adding them to the corresponding elements of any other row.**

Definition

The *main diagonal* of a matrix consists of all elements a_{ij} where $i = j$.

Definition

A matrix is in *triangular form* when all elements below the main diagonal are zero.

$$\begin{bmatrix} a_{11} & a_{12} & a_{13} & a_{14} & \cdots & a_{1n} \\ 0 & a_{22} & a_{23} & a_{24} & \cdots & a_{2n} \\ 0 & 0 & a_{33} & a_{34} & \cdots & a_{3n} \\ 0 & 0 & 0 & a_{44} & \cdots & a_{4n} \\ \cdot & & & & & \cdot \\ \cdot & & & & & \cdot \\ \cdot & & & & & \cdot \\ 0 & 0 & 0 & 0 & \cdots & a_{mn} \end{bmatrix}$$

We will now use the elementary row operations to reduce a matrix to an equivalent matrix in triangular form. This will be known as the *matrix method* (or gaussian elimination method) of solving a system of equations.

EXAMPLE   Solve the following system using the matrix method:

$$\begin{cases} 2x + 3y - z = 11 \\ x + 2y + z = 12 \\ 3x - y + 2z = 5 \end{cases}$$

The augmented matrix is

$$\begin{bmatrix} 2 & 3 & -1 & 11 \\ 1 & 2 & 1 & 12 \\ 3 & -1 & 2 & 5 \end{bmatrix}$$

If any variables were missing, we would place a zero in that position. Our goal is to use the elementary row operations to obtain a matrix of an equivalent system with zero entries in the a_{21}, a_{31}, and a_{32} positions. This might be accomplished by many different combinations of the elementary row operations. We will proceed as follows.

Multiply row two by -3 and add to row three.

$$\begin{bmatrix} 2 & 3 & -1 & 11 \\ 1 & 2 & 1 & 12 \\ 0 & -7 & -1 & -31 \end{bmatrix}$$

Multiply row two by -2.

$$\begin{bmatrix} 2 & 3 & -1 & 11 \\ -2 & -4 & -2 & -24 \\ 0 & -7 & -1 & -31 \end{bmatrix}$$

Add row one to row two.

$$\begin{bmatrix} 2 & 3 & -1 & 11 \\ 0 & -1 & -3 & -13 \\ 0 & -7 & -1 & -31 \end{bmatrix}$$

4.4 MATRICES

Multiply row two by −7.

$$\begin{bmatrix} 2 & 3 & -1 & 11 \\ 0 & 7 & 21 & 91 \\ 0 & -7 & -1 & -31 \end{bmatrix}$$

Add row two to row three.

$$\begin{bmatrix} 2 & 3 & -1 & 11 \\ 0 & 7 & 21 & 91 \\ 0 & 0 & 20 & 60 \end{bmatrix}$$

Since we have used only elementary row operations, the final matrix is the augmented matrix of a system that is equivalent to our original system. This equivalent system is

$$\begin{cases} 2x + 3y - z = 11 \\ 7y + 21z = 91 \\ 20z = 60 \end{cases}$$

Solving the third equation for z gives $z = 3$. Substituting $z = 3$ in the second equation gives $y = 4$. Substituting $z = 3$ and $y = 4$ in the first equation gives $x = 1$. Our solution is $(1,4,3)$, and this solution checks in the original system.

A triangular matrix could be reached by a different combination of the elementary row operations, but the solution would not be changed. Looking ahead carefully toward the desired form can result in arriving at the triangular form in a minimum number of steps.

Exercise 4.4.1

Solve the following systems of equations by using the matrix method:

1. $\begin{cases} 2x + 3y = 5 \\ 5x + y = 19 \end{cases}$
2. $\begin{cases} 3x + 5y = -15 \\ 2x + 7y = -10 \end{cases}$
3. $\begin{cases} 2x + y = -3 \\ 5x - y = 24 \end{cases}$

4. $\begin{cases} x + y - z = -2 \\ 2x - y + 3z = 19 \\ 4x + 3y - z = 5 \end{cases}$
5. $\begin{cases} 5x - y + 4z = 5 \\ 6x + y - 5z = 17 \\ 2x - 3y + z = -11 \end{cases}$
6. $\begin{cases} 4x + 3y + 5z = 2 \\ 2y + 7z = 16 \\ 2x - y = 6 \end{cases}$

7. $\begin{cases} 6x - y + z = 9 \\ 2x + 3z = 16 \\ 4x + 7y + 5z = 20 \end{cases}$ 8. $\begin{cases} 3x + 2y = 44 \\ 4y + 3z = 19 \\ 2x + 3z = -5 \end{cases}$

9. A woman made two investments, one at 5% interest and the other at 8%. Her total interest for the year was $1,700. The interest at 8% was $200 more than twice the interest at 5%. How much did she have invested at each rate?

10. The sum of three numbers is 58. Twice the first number added to the sum of the second and third is 71. If the first number is added to four times the second number and the sum is decreased by three times the third number, the result is 18. Find the numbers.

4.5 EVALUATING SECOND- AND THIRD-ORDER DETERMINANTS

Goals

Upon completion of this section you should be able to:
1. Find the determinant of a second-order matrix.
2. Find the determinant of a third-order matrix.

* *

Every square matrix (same number of rows and columns) has associated with it a unique real number called the determinant of the matrix. In the following sections we will discuss the determinant of two-by-two and three-by-three matrices and their uses in solving systems of linear equations.

Definition

The *determinant* of the second-order matrix

$$\begin{bmatrix} a & c \\ b & d \end{bmatrix} \text{ represented by } \begin{vmatrix} a & c \\ b & d \end{vmatrix}$$

is the real number $ad - bc$.

EXAMPLE

Evaluate $\begin{vmatrix} 2 & 3 \\ 1 & 4 \end{vmatrix}$.

$\begin{vmatrix} 2 & 3 \\ 1 & 4 \end{vmatrix} = (2)(4) - (1)(3) = 8 - 3 = 5$

EXAMPLE Evaluate $\begin{vmatrix} 1 & -2 \\ 3 & 2 \end{vmatrix}$.

$$\begin{vmatrix} 1 & -2 \\ 3 & 2 \end{vmatrix} = (1)(2) - (3)(-2) = 2 + 6 = 8$$

EXAMPLE Evaluate $\begin{vmatrix} 4 & 7 \\ 3 & 2 \end{vmatrix}$.

$$\begin{vmatrix} 4 & 7 \\ 3 & 2 \end{vmatrix} = (4)(2) - (3)(7) = 8 - 21 = -13$$

Exercise 4.5.1

Evaluate the following:

1. $\begin{vmatrix} 2 & 1 \\ 3 & 5 \end{vmatrix}$
2. $\begin{vmatrix} 3 & -2 \\ 1 & 2 \end{vmatrix}$
3. $\begin{vmatrix} 1 & 2 \\ 4 & 5 \end{vmatrix}$

4. $\begin{vmatrix} 2 & 4 \\ 3 & 6 \end{vmatrix}$
5. $\begin{vmatrix} 5 & -2 \\ 3 & 1 \end{vmatrix}$
6. $\begin{vmatrix} -2 & -3 \\ 5 & 4 \end{vmatrix}$

7. $\begin{vmatrix} 4 & 2 \\ -3 & -2 \end{vmatrix}$
8. $\begin{vmatrix} 4 & 0 \\ 6 & -3 \end{vmatrix}$
9. $\begin{vmatrix} -15 & -5 \\ 3 & 1 \end{vmatrix}$

10. $\begin{vmatrix} 3 & 7 \\ 0 & -2 \end{vmatrix}$
11. $\begin{vmatrix} 5 & 9 \\ 4 & -13 \end{vmatrix}$
12. $\begin{vmatrix} 0 & 0 \\ 3 & -6 \end{vmatrix}$

13. $\begin{vmatrix} 4 & 0 \\ 9 & 0 \end{vmatrix}$
14. $\begin{vmatrix} 8 & -4 \\ 6 & -3 \end{vmatrix}$
15. $\begin{vmatrix} 0 & -6 \\ 9 & 0 \end{vmatrix}$

The determinant of a third-order matrix can be found in more than one way, but we will use a general method that can be used to evaluate the determinant of a square matrix of any order.

Definition The *minor* of an element of a matrix is the determinant of the matrix that results when the row and column in which the element appears are eliminated.

EXAMPLE

In the matrix

$$\begin{bmatrix} 3 & 1 & 4 \\ 2 & 6 & 5 \\ 3 & -1 & 2 \end{bmatrix}$$

the minor of 1 (first row, second column) is obtained by eliminating the first row and second column

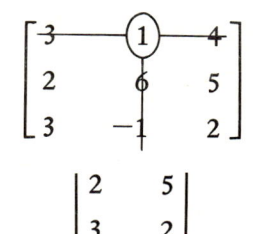

$$\begin{bmatrix} \cancel{3} & \cancel{1} & \cancel{4} \\ 2 & \cancel{6} & 5 \\ 3 & \cancel{-1} & 2 \end{bmatrix}$$

giving $\begin{vmatrix} 2 & 5 \\ 3 & 2 \end{vmatrix}$

and the minor of 5 (second row, third column) is

$$\begin{vmatrix} 3 & 1 \\ 3 & -1 \end{vmatrix}$$

Definition

The *cofactor* of an element of a matrix is its minor multiplied by $(-1)^{i+j}$, where i is the number of the row and j is the number of the column of the element.

Notice that $(-1)^{i+j}$ will be positive if $(i+j)$ is an even number and negative if $(i+j)$ is an odd number.

Definition

The *determinant of a third-order matrix* is the sum of the products of each element of a row or column and its cofactor.
For instance, in

$$\begin{bmatrix} a_{11} & a_{12} & a_{13} \\ a_{21} & a_{22} & a_{23} \\ a_{31} & a_{32} & a_{33} \end{bmatrix}$$

the determinant is

$$a_{11} \cdot \text{cofactor } a_{11} + a_{21} \cdot \text{cofactor } a_{21} + a_{31} \cdot \text{cofactor } a_{31}$$

4.5 EVALUATING SECOND- AND THIRD-ORDER DETERMINANTS

EXAMPLE Evaluate the following: $\begin{vmatrix} 2 & 1 & 3 \\ 3 & -5 & 4 \\ 5 & 0 & 2 \end{vmatrix}$

When we apply the definition to column one we get

$\begin{vmatrix} 2 & 1 & 3 \\ 3 & -5 & 4 \\ 5 & 0 & 2 \end{vmatrix} = 2(-1)^{1+1} \begin{vmatrix} -5 & 4 \\ 0 & 2 \end{vmatrix} + 3(-1)^{2+1} \begin{vmatrix} 1 & 3 \\ 0 & 2 \end{vmatrix}$

$\qquad\qquad\qquad\qquad\qquad + 5(-1)^{3+1} \begin{vmatrix} 1 & 3 \\ -5 & 4 \end{vmatrix}$

$\qquad = 2(1)(-10) + 3(-1)(2) + 5(1)(19)$
$\qquad = -20 - 6 + 95$
$\qquad = 69$

If we instead apply the definition to row three, we get

$\begin{vmatrix} 2 & 1 & 3 \\ 3 & -5 & 4 \\ 5 & 0 & 2 \end{vmatrix} = 5(-1)^{3+1} \begin{vmatrix} 1 & 3 \\ -5 & 4 \end{vmatrix} + 0(-1)^{3+2} \begin{vmatrix} 2 & 3 \\ 3 & 4 \end{vmatrix}$

$\qquad\qquad\qquad\qquad\qquad + 2(-1)^{3+3} \begin{vmatrix} 2 & 1 \\ 3 & -5 \end{vmatrix}$

$\qquad = 5(1)(19) + 0(-1)(-1) + 2(1)(-13)$
$\qquad = 95 + 0 - 26$
$\qquad = 69$

The value of a determinant is a constant and can be found by applying the definition to any row or column. Notice that choosing a row or column containing a zero cuts the amount of computation necessary.

EXAMPLE Evaluate the following: $\begin{vmatrix} 3 & 0 & 5 \\ 2 & 1 & 6 \\ -4 & 3 & -2 \end{vmatrix}$

It is wise to choose either the first row or the second column since the zero will simplify the work. Using the second column gives

$$\begin{vmatrix} 3 & 0 & 5 \\ 2 & 1 & 6 \\ -4 & 3 & -2 \end{vmatrix} = 1(-1)^{2+2} \begin{vmatrix} 3 & 5 \\ -4 & -2 \end{vmatrix} + 3(-1)^{3+2} \begin{vmatrix} 3 & 5 \\ 2 & 6 \end{vmatrix}$$

$$= 1(1)(14) + 3(-1)(8)$$
$$= 14 - 24$$
$$= -10$$

Exercise 4.5.2

Evaluate the following:

1. $\begin{vmatrix} 1 & 2 & 5 \\ 3 & 1 & 4 \\ 2 & 0 & 3 \end{vmatrix}$ 2. $\begin{vmatrix} 3 & 4 & 1 \\ 1 & 0 & -2 \\ 5 & -3 & 1 \end{vmatrix}$ 3. $\begin{vmatrix} 4 & -2 & 1 \\ -2 & 3 & 5 \\ 1 & 0 & 2 \end{vmatrix}$

4. $\begin{vmatrix} 6 & -1 & 5 \\ 1 & 0 & 4 \\ -5 & -2 & 3 \end{vmatrix}$ 5. $\begin{vmatrix} -1 & -2 & 4 \\ 3 & 1 & -1 \\ 5 & 6 & 5 \end{vmatrix}$ 6. $\begin{vmatrix} 4 & 0 & 1 \\ -2 & 1 & 0 \\ 7 & 3 & 5 \end{vmatrix}$

7. $\begin{vmatrix} 3 & 5 & -3 \\ 4 & 1 & -2 \\ -2 & 6 & -4 \end{vmatrix}$ 8. $\begin{vmatrix} 0 & 0 & 0 \\ -2 & 3 & -1 \\ 4 & 8 & -2 \end{vmatrix}$ 9. $\begin{vmatrix} 4 & -2 & 0 \\ 3 & 4 & 0 \\ 1 & 3 & 0 \end{vmatrix}$

10. $\begin{vmatrix} -1 & 3 & 5 \\ -2 & 1 & -1 \\ 4 & 6 & 5 \end{vmatrix}$

11. Show that the equation of a line containing the points (x_1, y_1) and (x_2, y_2) may be expressed as

$$\begin{vmatrix} x & y & 1 \\ x_1 & y_1 & 1 \\ x_2 & y_2 & 1 \end{vmatrix} = 0$$

(*Hint*: Evaluate the determinant on the left. Then show that the two-point form of the equation of a line is equivalent to this form.)

12. Use the determinant form discussed in Problem 11 to find the equation of a line through the points (3,1) and (2,3).

4.6 SOLVING SYSTEMS OF FIRST-DEGREE EQUATIONS BY DETERMINANTS (TWO EQUATIONS WITH TWO VARIABLES)

Goals

Upon completion of this section you should be able to:

1. Solve a system of two first-degree equations by the determinant method.
2. Use determinants to classify a system of two first-degree equations as independent, inconsistent, or dependent.

* *

The *standard form* of a system of two linear equations is

$$\begin{cases} ax + cy = e \\ bx + dy = f \end{cases}$$

If we solve this system, either by substitution or addition, the solution will be

$$x = \frac{de - cf}{ad - bc}$$

$$y = \frac{af - be}{ad - bc}$$

From the definition of the value of the determinant of a second-order matrix, we see that these values for x and y could be expressed as

$$x = \frac{\begin{vmatrix} e & c \\ f & d \end{vmatrix}}{\begin{vmatrix} a & c \\ b & d \end{vmatrix}}$$

$$y = \frac{\begin{vmatrix} a & e \\ b & f \end{vmatrix}}{\begin{vmatrix} a & c \\ b & d \end{vmatrix}}$$

This gives us a method of solving a system of two linear equations by determinants. This method is often referred to as *Cramer's Rule*. In a sense this is a formula and, as all formulas, must come from the standard form. If an equation is not in standard form, then it must first be put in standard form before this method will apply.

EXAMPLE

Solve by determinants $\begin{cases} 3x - y = 5 \\ x + y = 3 \end{cases}$.

The equations are already in standard form so

$$x = \frac{\begin{vmatrix} 5 & -1 \\ 3 & 1 \end{vmatrix}}{\begin{vmatrix} 3 & -1 \\ 1 & 1 \end{vmatrix}} = \frac{8}{4} = 2$$

$$y = \frac{\begin{vmatrix} 3 & 5 \\ 1 & 3 \end{vmatrix}}{\begin{vmatrix} 3 & -1 \\ 1 & 1 \end{vmatrix}} = \frac{4}{4} = 1$$

The solution is (2,1).

Some special patterns can be noted which make it easy to set up the determinants involved.

When the equations are in standard form:

1. The denominator of each variable is the determinant of the matrix of coefficients.
2. The numerator is the same as the denominator *except* that the column of the variable being found is replaced by the column of constants.

EXAMPLE

Solve by determinants $\begin{cases} 4x + 3y = 2 \\ 3x + 4y = 5 \end{cases}$.

$$x = \frac{\begin{vmatrix} 2 & 3 \\ 5 & 4 \end{vmatrix}}{\begin{vmatrix} 4 & 3 \\ 3 & 4 \end{vmatrix}} = \frac{-7}{7} = -1$$

4.6 SOLVING FIRST-DEGREE EQUATIONS BY DETERMINANTS

$$y = \frac{\begin{vmatrix} 4 & 2 \\ 3 & 5 \end{vmatrix}}{\begin{vmatrix} 4 & 3 \\ 3 & 4 \end{vmatrix}} = \frac{14}{7} = 2$$

The solution is (−1,2).

EXAMPLE

Solve by determinants

$$\begin{cases} y = 2x - 1 \\ \dfrac{x+1}{2} + \dfrac{3y}{5} = \dfrac{8}{5} \end{cases}$$

Put the equation in a standard form by rewriting the first equation and clearing the fractions in the second. The first equation becomes

$$2x - y = 1$$

Multiplying both sides of the second equation by the least common denominator (10) and rearranging gives

$$5x + 6y = 11$$

We now have the system

$$\begin{cases} 2x - y = 1 \\ 5x + 6y = 11 \end{cases}$$

Hence

$$x = \frac{\begin{vmatrix} 1 & -1 \\ 11 & 6 \end{vmatrix}}{\begin{vmatrix} 2 & -1 \\ 5 & 6 \end{vmatrix}} = \frac{17}{17} = 1$$

$$y = \frac{\begin{vmatrix} 2 & 1 \\ 5 & 11 \end{vmatrix}}{\begin{vmatrix} 2 & -1 \\ 5 & 6 \end{vmatrix}} = \frac{17}{17} = 1$$

The solution is (1,1).

EXAMPLE Solve by determinants $\begin{cases} x - 2y = 3 \\ 3x - 6y = 9 \end{cases}$.

$$x = \frac{\begin{vmatrix} 3 & -2 \\ 9 & -6 \end{vmatrix}}{\begin{vmatrix} 1 & -2 \\ 3 & -6 \end{vmatrix}} = \frac{-18 + 18}{-6 + 6} = \frac{0}{0}$$

$$y = \frac{\begin{vmatrix} 1 & 3 \\ 3 & 9 \end{vmatrix}}{\begin{vmatrix} 1 & -2 \\ 3 & -6 \end{vmatrix}} = \frac{9 - 9}{-6 + 6} = \frac{0}{0}$$

Notice that *both* numerator and denominator are zero. These, of course, are undefined quantities, and we seem to be at an impasse.

Try solving the system by the addition method. Do you see that the equations are dependent?

We have thus discovered a characteristic of a system of dependent equations when solving by determinants: *if the determinants in both numerator and denominator are zero, the equations are dependent.*

EXAMPLE Solve by determinants $\begin{cases} 2x + 4y = 5 \\ x + 2y = 6 \end{cases}$.

$$x = \frac{\begin{vmatrix} 5 & 4 \\ 6 & 2 \end{vmatrix}}{\begin{vmatrix} 2 & 4 \\ 1 & 2 \end{vmatrix}} = \frac{10 - 24}{4 - 4} = \frac{-14}{0}$$

$$y = \frac{\begin{vmatrix} 2 & 5 \\ 1 & 6 \end{vmatrix}}{\begin{vmatrix} 2 & 4 \\ 1 & 2 \end{vmatrix}} = \frac{12 - 5}{4 - 4} = \frac{7}{0}$$

4.6 SOLVING FIRST-DEGREE EQUATIONS BY DETERMINANTS

Here again we see meaningless values for x and y since the denominators are zero.

Solve the system by the addition method. Do you see that the equations are inconsistent?

Thus when solving a system by determinants: *if the determinant of the matrix of coefficients is zero but the other determinants (numerators) are not, the equations are inconsistent.*

Exercise 4.6.1

Solve the following systems by determinants:

1. $\begin{cases} x + y = 3 \\ 2x - y = 0 \end{cases}$
2. $\begin{cases} 2x + y = 7 \\ x + 2y = 11 \end{cases}$
3. $\begin{cases} x - y = -1 \\ 2x - y = 1 \end{cases}$

4. $\begin{cases} 2x + 3y = 10 \\ x - y = -5 \end{cases}$
5. $\begin{cases} 3x + y = 2 \\ 6x + 2y = 4 \end{cases}$
6. $\begin{cases} 3x + 2y = -8 \\ 2x - 5y = 1 \end{cases}$

7. $\begin{cases} 2x - 3y = 6 \\ x + 2y = 3 \end{cases}$
8. $\begin{cases} 2x + 8y = 1 \\ x + 4y = 3 \end{cases}$
9. $\begin{cases} x + 4y = 1 \\ 2x - 3y = 13 \end{cases}$

10. $\begin{cases} 2x + y = 0 \\ 4x + 3y = 6 \end{cases}$
11. $\begin{cases} x = 2y - 4 \\ 3y = 5 + x \end{cases}$
12. $\begin{cases} 2x - 3y = 5 \\ \frac{4}{3}x - 2y = \frac{10}{3} \end{cases}$

13. $\begin{cases} x - 2y = 6 \\ \frac{x-1}{3} + \frac{y}{2} = \frac{1}{2} \end{cases}$
14. $\begin{cases} y = x + 3 \\ x = \frac{2y}{3} - 2 \end{cases}$
15. $\begin{cases} \frac{x+1}{5} + \frac{y-3}{2} = 1 \\ y = -5x \end{cases}$

16. The sum of Tom's age and twice Kathy's age is 57. Three times Kathy's age, decreased by twice Tom's age, is 12. Find their ages.

4.7 SOLVING SYSTEMS OF FIRST-DEGREE EQUATIONS BY DETERMINANTS (THREE EQUATIONS WITH THREE VARIABLES)

Goals

Upon completion of this section you should be able to:

1. Solve a system of three first-degree equations in three variables by determinants.

2. Use determinants to classify a system of three first-degree equations in three variables as independent, inconsistent, or dependent.

* *

A system of three first-degree equations in three variables can be solved by following the same pattern used for two equations with two variables in the last section (Cramer's Rule).

If the equations are in standard form, then

1. The denominator of each variable is the same and is the determinant of the matrix of coefficients of the variables as they appear.
2. The numerator is the same as the denominator *except* that the column of the variable being found is replaced by the column of constants.

EXAMPLE   Solve by determinants

$$\begin{cases} 2x - 3y + z = 7 \\ x + y + z = 2 \\ 3x + 3y - z = -2 \end{cases}$$

The equations are in standard form, so we set up the determinants to find x, y, and z as just discussed.

$$x = \frac{\begin{vmatrix} 7 & -3 & 1 \\ 2 & 1 & 1 \\ -2 & 3 & -1 \end{vmatrix}}{\begin{vmatrix} 2 & -3 & 1 \\ 1 & 1 & 1 \\ 3 & 3 & -1 \end{vmatrix}} = \frac{-20}{-20} = 1$$

$$y = \frac{\begin{vmatrix} 2 & 7 & 1 \\ 1 & 2 & 1 \\ 3 & -2 & -1 \end{vmatrix}}{\begin{vmatrix} 2 & -3 & 1 \\ 1 & 1 & 1 \\ 3 & 3 & -1 \end{vmatrix}} = \frac{20}{-20} = -1$$

4.7 SOLVING FIRST-DEGREE EQUATIONS BY DETERMINANTS

$$z = \frac{\begin{vmatrix} 2 & -3 & 7 \\ 1 & 1 & 2 \\ 3 & 3 & -2 \end{vmatrix}}{\begin{vmatrix} 2 & -3 & 1 \\ 1 & 1 & 1 \\ 3 & 3 & -1 \end{vmatrix}} = \frac{-40}{-20} = 2$$

Thus the solution is $(1,-1,2)$.

Note that the denominator need not be evaluated each time since it is the same. Also, it is possible to find x and y and then substitute these values in one of the equations to find the z. Thus it is possible to solve the system by evaluating three third-order determinants and making one substitution.

CAUTION: Since the denominator is the same for each, we should be extremely careful when it is evaluated!

Should the value of the denominator be zero, two possibilities exist.
1. If the numerators of all the variables are zero, the system is dependent.
2. If at the least one of the variables has a nonzero numerator, the system is inconsistent.

Exercise 4.7.1.

Solve the following systems by determinants:

1. $\begin{cases} x + y + z = 6 \\ 2x - y + 2z = 6 \\ 3x + 2y - z = 4 \end{cases}$

2. $\begin{cases} x - y + 2z = 3 \\ 3x + y + z = -1 \\ 2x - 3y + 5z = 8 \end{cases}$

3. $\begin{cases} x + 2y - z = 13 \\ 2x - y - 2z = 11 \\ 3x + y + z = 4 \end{cases}$

4. $\begin{cases} 2x + y - z = -2 \\ 3x + 2y + z = -4 \\ x - y + 3z = 13 \end{cases}$

5. $\begin{cases} 3x - y + z = -10 \\ 2x + 3y - 2z = -3 \\ x - 5y + 3z = -8 \end{cases}$

6. $\begin{cases} x + y = -2 \\ 3x - z = 1 \\ 2x + y + z = 1 \end{cases}$

7. $\begin{cases} x + 2y - 3z = -15 \\ 3x - y + z = 9 \\ 2x + 3y - z = -8 \end{cases}$

8. $\begin{cases} x - 3z = -15 \\ x - 2y = 2 \\ y + z = 4 \end{cases}$

9. The sum of three numbers is zero. Twice the first number, added to the second, is 11 less than the third. The third number is 17 more than the second. Find the numbers.

10. A couple buys 11 gallons of three different kinds of paint, some at $8.00 a gallon, some at $7.00 a gallon, and some at $5.00 a gallon. They have twice as many gallons of $5.00 paint as they do of $7.00 paint. If the total bill for the paint is $58.00, how many gallons of $5.00 paint did they buy?

4.8 SOLVING SYSTEMS OF INEQUALITIES IN TWO VARIABLES BY GRAPHING

Goals

Upon completion of this section you should be able to:

1. Sketch the graph of an inequality in two variables.
2. Use the graphical method to solve a system of inequalities.

* *

Systems of inequalities and their solutions are the basis for a branch of mathematics called *linear programming*. Linear programming is very useful in business as well as some scientific fields, and the student should have a background in solving systems of inequalities.

In previous sections we have solved inequalities in one variable. In this section we will graph inequalities in two variables and solve systems of inequalities in two variables by graphical methods.

There is a rather obvious relationship between the graphs of equations and inequalities. For instance, the graph of $2x + 3y = 7$ is a straight line while the graph of $2x + 3y > 7$ is the region above the line and the graph of $2x + 3y < 7$ is the region below the line. The graph of $x^2 + y^2 = 36$ is a circle and the graph of $x^2 + y^2 < 36$ is the interior of the circle while the graph of $x^2 + y^2 > 36$ is the exterior of the circle. Our method of graphing inequalities will then involve two steps:

1. Graph the equation found by replacing the inequality sign with an equal sign.
2. Determine which region is represented by the inequality.

EXAMPLE

Graph $2x - y < 8$. The first step is to graph the line $2x - y = 8$.

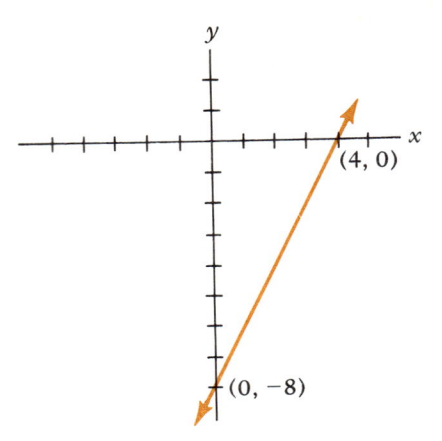

Graph of $2x - y = 8$

Our inequality $2x - y < 8$ will either be the region above the line or below the line. To determine which region, it is only necessary that we choose a specific point that is obviously not on the line and see if it is in the region that is the solution set for the inequality. The origin $(0,0)$ is always a good choice if the curve does not go through that point. In this case, substituting $(0,0)$ into $2x - y < 8$ gives $0 < 8$ which is a true statement. Therefore, the region above the line containing the origin is the solution set for $2x - y < 8$, and the graph is

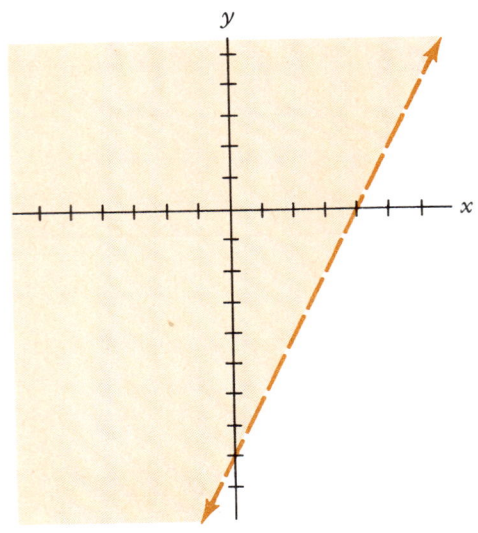

Graph of $2x - y < 8$

Notice that the region does *not* include the line $2x - y = 8$, and we draw it as a broken (or dashed) line to indicate this fact. If (0,0) had given us a false statement, the region not containing (0,0) would have been the solution set.

EXAMPLE   Graph $x^2 + y^2 - 6x - 10y \geq 2$. First we will complete the square and get the equation $x^2 + y^2 - 6x - 10y = 2$ into standard form.

$$(x - 3)^2 + (y - 5)^2 = 36$$

We now see that we have the equation of a circle with its center at (3,5) and a radius of 6.

Our inequality $x^2 + y^2 - 6x - 10y \geq 2$ is either the interior or exterior of the circle together with the circle itself. We may use the center (3,5) as a checkpoint. Substituting (3,5) into the inequality gives $-34 \geq 2$ which is a false statement. Therefore, the region containing (3,5) is not in the solution set. As mentioned, since we have $\geq$ as our relation, the circle itself is also a part of the solution set and is drawn as an unbroken curve to indicate this fact. The graph is

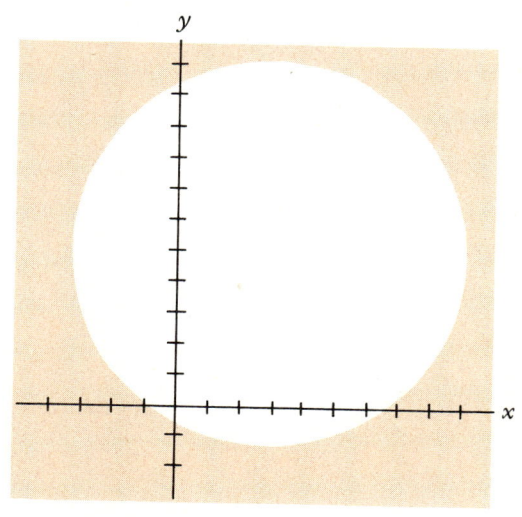

Graph of $x^2 + y^2 - 6x - 10y \geq 2$

Exercise 4.8.1

Graph the solution set for each of the following inequalities:

1. $x + y > 3$
2. $2x - 3y \leq 0$

4.8 SOLVING INEQUALITIES IN TWO VARIABLES BY GRAPHING

3. $x^2 + y^2 - 2x - 6y < -6$
4. $x^2 + y^2 + 8y \geq 0$
5. $9x^2 + 18x + 16y^2 - 64y \leq 71$
6. $x^2 - 8x - 8y \geq -40$
7. $y^2 - 4x^2 > 16$
8. $y^2 - 12x > 12$

When we graph a system of two or more inequalities on the same coordinate axes, the intersection of the solution sets will be the solution set of the system.

EXAMPLE

Graph the solution set of the system $\begin{cases} x - y \geq 6 \\ 2x + 3y \leq 5 \end{cases}$.

We first graph the lines $x - y = 6$ and $2x + 3y = 5$. Neither line goes through (0,0) and substituting these coordinates into $x - y \geq 6$ gives $0 \geq 6$, which is a false statement. The solution set for $x - y \geq 6$ is, therefore, the region not containing the origin.

Substituting (0,0) into $2x + 3y \leq 5$ gives $0 \leq 5$ which implies that (0,0) is in the solution set of $2x + 3y \leq 5$. The graph is

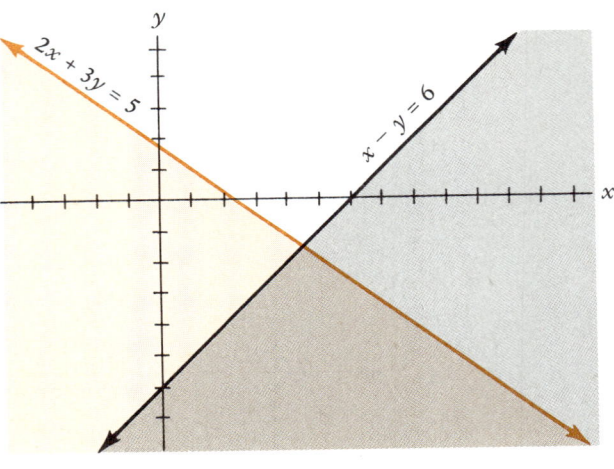

Graph of $\begin{cases} x - y \geq 6 \\ 2x + 3y \leq 5 \end{cases}$

The double-shaded portion of the plane is the solution set of the system of the two inequalities.

EXAMPLE Graph the system

$$\begin{cases} x + y \geq 3 \\ x + y \leq 5 \\ x \geq 0 \\ y \geq 0 \end{cases}$$

Graphing the equations $x + y = 3$, $x + y = 5$, $x = 0$, and $y = 0$ gives us two parallel lines and the two axes. Substitution of coordinates in the original inequalities will then lead to the following graph of the solution set of the system.

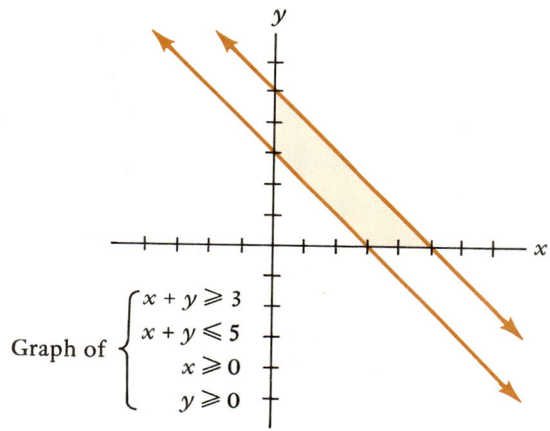

Graph of $\begin{cases} x + y \geq 3 \\ x + y \leq 5 \\ x \geq 0 \\ y \geq 0 \end{cases}$

Notice that this system has a solution that is the trapezoidal region having vertices at (0,3), (0,5), (3,0), and (5,0).

Sometimes a linear equation is used in conjunction with inequalities involving the same variables. This occurs in the branch of mathematics known as *linear programming*. The following example illustrates the process.

EXAMPLE A small manufacturing company produces two models of fishing reels. Model A sells for $25 and model B sells for $40. The company cannot sell more than 100 reels a week. It takes 1 hour to produce a model A reel and 3 hours to produce a model B reel. The total number of labor hours available in a week is 150. How many reels of each model should the company produce in a week to maximize income?

First we shall let

x = number of model A reels to be produced

y = number of model B reels to be produced

Then since the total number of reels produced cannot exceed 100 we have

$$x + y \leq 100$$

It would take a total of x hours to produce the model A reels and $3y$ hours to produce the model B reels. Thus

$$x + 3y \leq 150$$

We therefore have the following system of inequalities

$$\begin{cases} x + y \leq 100 \\ x + 3y \leq 150 \\ x \geq 0 \\ y \geq 0 \end{cases}$$

The graph of the system follows

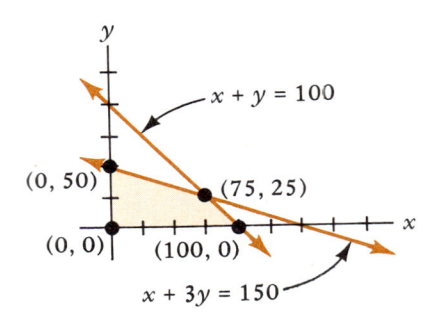

Now the income for the company can be expressed as

$$I = 25x + 40y$$

This expression has certain restrictions or *contraints* on the values of x and y as provided for by the given system of inequalities. We wish to find which values for x and y under the given constraints will produce a maximum income.

At this point we must introduce an important theorem from linear programming. It states that when we have a region, such as in this example: *the maximum and minimum values (if they exist) for the linear expression will be found at the vertices of the region.*

Thus if there is a maximum income, it must occur at one of the vertices. We therefore try the coordinates for each vertex in the expression for income.

$$(0,0): \quad I = 25(0) + 40(0) = \$0$$
$$(0,50): \quad I = 25(0) + 40(50) = \$2{,}000$$
$$(75,25): \quad I = 25(75) + 40(25) = \$2{,}875$$
$$(100,0): \quad I = 25(100) + 40(0) = \$2{,}500$$

We see that the vertex (75,25) produces the maximum income. Therefore, the company should produce 75 model A reels and 25 model B reels.

Exercise 4.8.2

Graph the solution sets of the following systems:

1. $\begin{cases} x + y \leqslant 2 \\ 2x - y \geqslant 1 \end{cases}$

2. $\begin{cases} 3x + y > -9 \\ x - 2y < 4 \end{cases}$

3. $\begin{cases} x - y \geqslant -1 \\ y \leqslant x^2 - 2x + 1 \end{cases}$

4. $\begin{cases} 2x + y > 8 \\ x^2 + y^2 - 6x - 2y \leqslant -1 \end{cases}$

5. $\begin{cases} y \geqslant 0 \\ x \geqslant -1 \\ x + y \geqslant 1 \end{cases}$

6. $\begin{cases} 3x + y \leqslant 0 \\ 2x - y \geqslant -5 \\ y \geqslant 0 \end{cases}$

7. $\begin{cases} x^2 + y^2 < 25 \\ x \geqslant -2 \\ y \geqslant 3 \end{cases}$

8. $\begin{cases} 3x - 4y \geqslant -12 \\ x + y \geqslant 1 \\ 3x + 4y \leqslant 12 \\ 2x - y \leqslant 2 \end{cases}$

9. Two numbers are such that the first number is not less than 10 and the sum of twice the first number and three times the second number is not more than 60. Also twice the first number decreased by the second number is not greater

than 40. Find the two numbers satisfying these constraints such that their sum will be minimum. Find the two numbers whose sum will be maximum.

10. A company wishes to purchase advertising time on a television network. The network offers a package containing one-half-minute commercials and one-minute commercials. The company must purchase a total of at least 100 minutes of air time. The number of one-minute commercials must be at least one and one-half that of the half-minute commercials. The number of one-minute commercials cannot exceed 125. If each half-minute commercial costs $30,000 and each one-minute commercial costs $50,000, how many of each should the company buy to minimize its expenses?

CHAPTER REVIEW

Solve the following systems by graphing:

1. $\begin{cases} x + y = -1 \\ 2x - y = 4 \end{cases}$
2. $\begin{cases} x - y = 3 \\ x + 2y = -6 \end{cases}$
3. $\begin{cases} 2x + y = -3 \\ x - 2y = -4 \end{cases}$

4. $\begin{cases} 3x - y = 7 \\ 2x + y = 8 \end{cases}$
5. $\begin{cases} x + y = -1 \\ y = x^2 - 1 \end{cases}$

Classify each of the following systems as independent, inconsistent, or dependent. If the system is independent, find its solution.

6. $\begin{cases} x + y = 1 \\ 2x - y = 5 \end{cases}$
7. $\begin{cases} 2x - y = 4 \\ 6x - 3y = 8 \end{cases}$
8. $\begin{cases} 2x + y = -5 \\ 3x + 2y = -10 \end{cases}$

9. $\begin{cases} 3x + 6y = 15 \\ 2x + 4y = 10 \end{cases}$
10. $\begin{cases} 2x + 3y = -5 \\ 5x - 2y = -22 \end{cases}$

Solve the following systems algebraically:

11. $\begin{cases} x + y - z = 4 \\ 2x + y + z = 3 \\ 2x + 2y + z = 5 \end{cases}$
12. $\begin{cases} x + y - z = -3 \\ x + y + z = 3 \\ 3x - y + z = 7 \end{cases}$
13. $\begin{cases} 2x - y + z = 5 \\ x + 2y - z = -2 \\ x + y - 2z = -5 \end{cases}$

14. $\begin{cases} 2x - 3y + z = 11 \\ x + y + 2z = 8 \\ x + 3y - z = -11 \end{cases}$
15. $\begin{cases} x + y = -2 \\ x - z = -1 \\ x + 2y - z = 1 \end{cases}$

Solve the following systems of equations using the matrix method:

16. $\begin{cases} x + 2y = 3 \\ 2x + 5y = 1 \end{cases}$

17. $\begin{cases} 2x + 3y = 6 \\ 3x - 4y = -8 \end{cases}$

18. $\begin{cases} x + y - z = 0 \\ 3x - y + z = 12 \\ 2x + 3y + 2z = 7 \end{cases}$

19. $\begin{cases} 3x - 2y + z = 4 \\ 4x + 3z = 12 \\ x - 5y + 2z = 8 \end{cases}$

20. $\begin{cases} 3x + z = -2 \\ 5y - 3z = 23 \\ 6x + 7y = 26 \end{cases}$

Evaluate the following determinants:

21. $\begin{vmatrix} 3 & -6 \\ 1 & 2 \end{vmatrix}$

22. $\begin{vmatrix} -4 & 16 \\ 0 & 3 \end{vmatrix}$

23. $\begin{vmatrix} 2 & -3 \\ 8 & -12 \end{vmatrix}$

24. $\begin{vmatrix} 2 & 1 & 4 \\ 1 & 0 & 2 \\ -3 & 1 & -3 \end{vmatrix}$

25. $\begin{vmatrix} -2 & 1 & 0 \\ 0 & 6 & -3 \\ 4 & 2 & -5 \end{vmatrix}$

Solve the following systems by determinants:

26. $\begin{cases} 2x + y = 5 \\ 3x - 2y = 11 \end{cases}$

27. $\begin{cases} x - y = 6 \\ 3x + 2y = -7 \end{cases}$

28. $\begin{cases} 2x + 3y = 2 \\ x - 5y = 14 \end{cases}$

29. $\begin{cases} x + 2y = -6 \\ 2x - 3y = 9 \end{cases}$

30. $\begin{cases} 2x + 5y = 0 \\ 3x - 2y = 19 \end{cases}$

Solve the following by determinants:

31. $\begin{cases} x + y + z = 2 \\ 2x - y + z = -1 \\ x - y + z = -2 \end{cases}$

32. $\begin{cases} x + y - z = -3 \\ x + y + z = 3 \\ 3x - y + z = 7 \end{cases}$

33. $\begin{cases} 2x - y + z = 5 \\ x + 2y - z = -2 \\ x + y - 2z = -5 \end{cases}$

34. $\begin{cases} 2x - 3y + z = 11 \\ x + y + 2z = 8 \\ x + 3y - z = -11 \end{cases}$

35. $\begin{cases} x + y = -2 \\ x - z = -1 \\ x + 2y - z = 1 \end{cases}$

36. A boat went a certain distance upstream and back to the starting point in 5 hours. The speed of the boat in still water is 15 miles per hour. The current of the stream is 3 miles per hour. How long did it take the boat to go upstream?

CHAPTER REVIEW 151

37. A person has 15 coins consisting of nickels, dimes, and quarters. The value of the coins is $2.20. If there are two more quarters than nickels, how many dimes are there?

Graph the solution sets of the following systems:

38. $\begin{cases} y - |x| \geqslant 0 \\ 2x - 3y \geqslant -6 \end{cases}$

39. $\begin{cases} x - y \leqslant -1 \\ 3x + 2y \leqslant 6 \end{cases}$

40. $\begin{cases} x + y < -1 \\ y \geqslant x^2 - 2x - 3 \end{cases}$

41. $\begin{cases} y \geqslant x^2 + 1 \\ x^2 + y^2 \leqslant 16 \\ y \geqslant 3 \end{cases}$

42. $\begin{cases} x - y \geqslant -4 \\ x + y \geqslant -2 \\ 2x + 3y \leqslant 6 \\ x \leqslant 0 \end{cases}$

43. A broker has a maximum of $12,000 to invest in two types of bonds. Bond A returns 8% and bond B returns 10% per year. The investment in bond B must not exceed 50% of the investment in bond A. How much should be invested at each rate in order to obtain a maximum profit?

PRACTICE TEST

1. Solve the following system by graphing:

$$\begin{cases} x + y = 4 \\ 2x - y = 5 \end{cases}$$

2. Solve the following system by algebraic methods:

$$\begin{cases} 2x + y = 2 \\ x - 3y = 15 \end{cases}$$

3. Solve the following system of equations using the matrix method. Show your work.

$$\begin{cases} 2x - y + z = 14 \\ 3x + 2y + 2z = 17 \\ x + 5y + 3z = 11 \end{cases}$$

4. Evalute the following determinants:

a. $\begin{vmatrix} 6 & 3 \\ 5 & -1 \end{vmatrix}$

b. $\begin{vmatrix} 4 & 0 & -3 \\ 1 & 2 & 5 \\ 3 & 1 & 6 \end{vmatrix}$

5. Solve the following system by determinants. Show your work.

$$\begin{cases} 2x + 3y = -2 \\ 3x + 2y = 7 \end{cases}$$

6. Solve the following system by any method. Show your work.

$$\begin{cases} x + 2y + z = 1 \\ 2x + 3y - z = 0 \\ x - 2y + 3z = 7 \end{cases}$$

7. Graph the solution set of the following system:

$$\begin{cases} y \geqslant x^2 + 2x - 3 \\ x - y \leqslant 2 \\ y \leqslant 0 \\ x \leqslant -1 \end{cases}$$

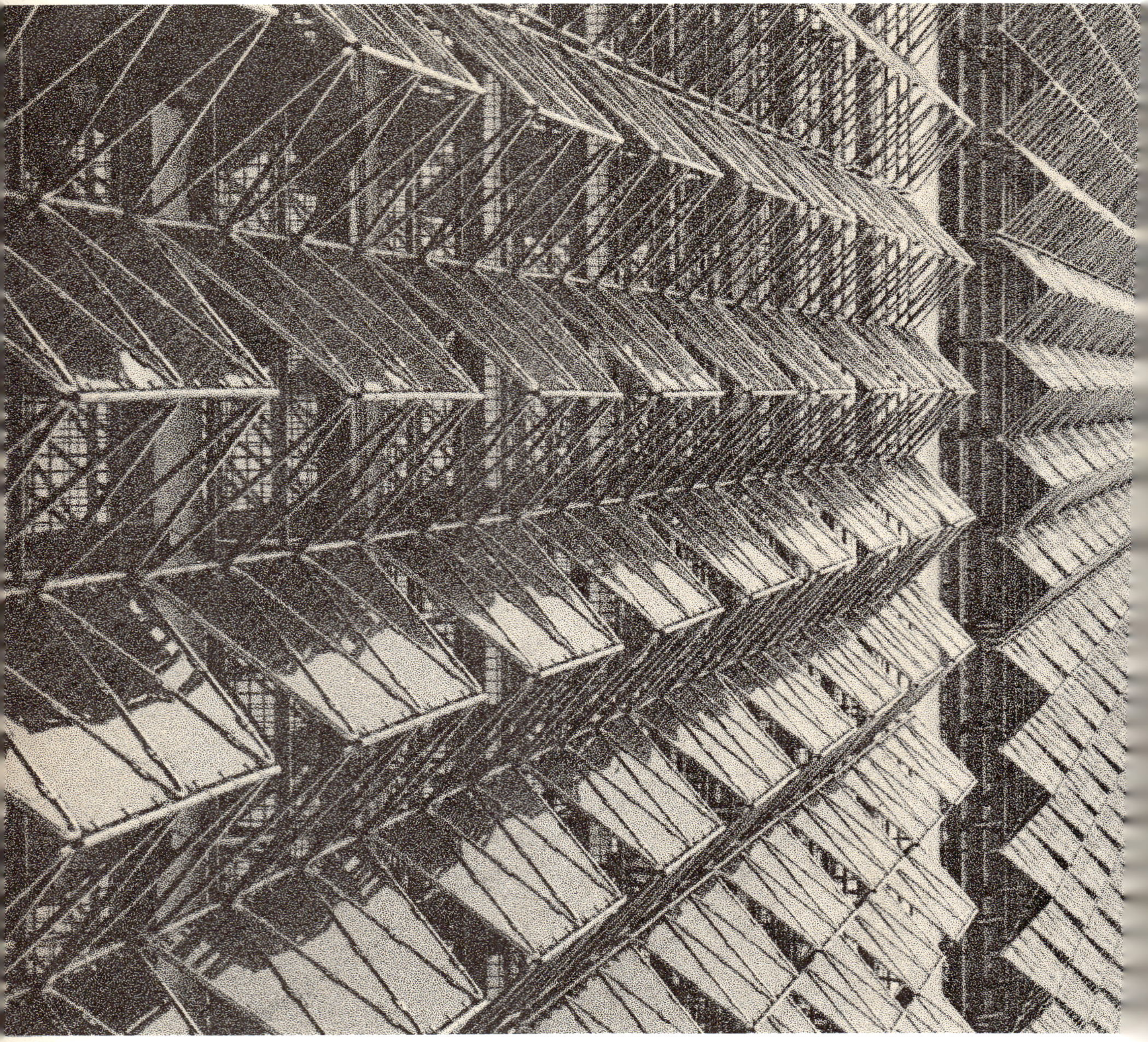

Logarithms

5

5

The general topic of logarithms has been a part of the study of mathematics since their invention by John Napier (1550–1617). At that time their main use was to simplify certain numerical computations. They were used as the basis for early computers—the slide rule is a classic example of the use of the logarithmic scale.

The advent of the modern electronic computer and the increasing popularity and availability of the hand-held calculator have diminished the need for the use of logarithms in arithmetic computations. However, the concept of a logarithm as a function is of eminent value in higher mathematics, and many branches of the sciences make use of them.

In this chapter we will give an overview of the computational use of logarithms (this is important for an understanding of the theory of logarithms) with major emphasis on the implications of the definition of the logarithmic function.

5.1 DEFINITION OF LOGARITHM

Goals

Upon completion of this section you should be able to:

1. Change an exponential statement of the form $x = b^y$ to logarithmic form.
2. Change a logarithmic statement of the form $\log_b x$ to exponential form.
3. Graph a logarithmic function.

* *

Definition

The *logarithm* of a number x to the base b is the exponent y to which the base b is raised to obtain x. (b can represent any positive number except 1.) In symbols we may write

$$\text{if } x = b^y, \text{ then } y = \log_b x$$

156   LOGARITHMS

We read $\log_b x$ as "the logarithm, base b, of x" or usually as "log, base b, of x." From the definition we see that

$$8 = 2^3 \quad \text{and} \quad \log_2 8 = 3$$

are equivalent statements. $8 = 2^3$ is the *exponential* form and $\log_2 8 = 3$ is the *logarithmic* form.

A sketch of the graph of the logarithmic curve might be helpful. We will sketch the graph of

$$y = \log_2 x$$

If we first change from logarithmic form to exponential form, the equation becomes

$$x = 2^y$$

We now set up a table of values choosing values for y and then finding the corresponding values for x.

x	$\frac{1}{8}$	$\frac{1}{4}$	$\frac{1}{2}$	1	2	4	8
y	-3	-2	-1	0	1	2	3

We then sketch the following curve.

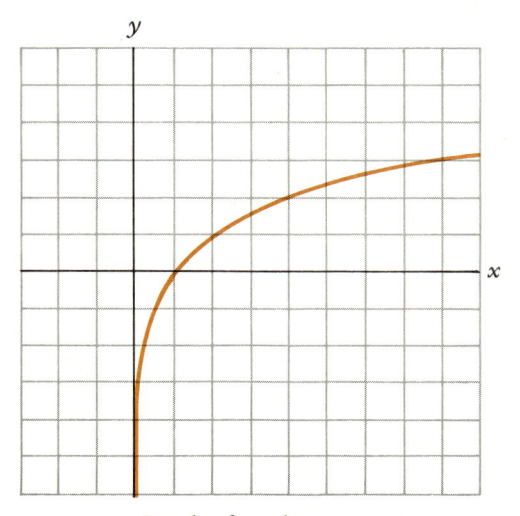

Graph of $y = \log_2 x$

5.1 DEFINITION OF LOGARITHM 157

Some observations from the graph regarding the properties of $y = \log_2 x$:

1. $y = \log_2 x$ is a function since each value of x is associated with only one y value.
2. Each y value is associated with only one x value.
3. The point (1,0) would be on the graph for the logarithm to any base.
4. Numbers less than or equal to zero have no logarithm. (The domain of the function is $(0, +\infty]$.)
5. Numbers greater than zero but less than 1 have negative logarithms.
6. Numbers greater than 1 have positive logarithms.
7. As the value of x increases, the value of y increases.
8. The graph is asymptotic to the y-axis.

These facts would hold true for any base greater than 1. A different base would change the rate at which the curve changes direction but not its general shape.

Exercise 5.1.1

Change to logarithmic form.

1. $3^2 = 9$
2. $5^2 = 25$
3. $5^3 = 125$
4. $49 = 7^2$
5. $27 = 3^3$
6. $64 = 2^6$
7. $\dfrac{1}{8} = 2^{-3}$
8. $3^0 = 1$
9. $\dfrac{1}{16} = 2^{-4}$
10. $3^1 = 3$

Change to exponential form.

11. $\log_2 4 = 2$
12. $\log_3 9 = 2$
13. $\log_4 16 = 2$
14. $\log_3 27 = 3$
15. $\log_3 81 = 4$
16. $\log_2 \dfrac{1}{2} = -1$
17. $\log_5 125 = 3$
18. $\log_2 32 = 5$

Find the value of the variable.

19. $\log_3 x = 4$
20. $\log_3 \dfrac{1}{3} = y$

LOGARITHMS

21. $\log_b 16 = 4$

22. $\log_2 \frac{1}{8} = y$

23. $\log_b 125 = 3$

24. $\log_2 \frac{1}{32} = y$

25. $\log_3 x = -2$

26. $\log_2 x = -4$

27. $\log_7 x = 2$

28. $\log_{10} 1 = y$

29. $\log_{10} 10 = y$

30. $\log_{10} 100 = y$

31. $\log_{10} 1,000 = y$

32. $\log_{10} \frac{1}{10} = y$

33. $\log_{10} \frac{1}{100} = y$

34. $\log_{10} \frac{1}{1,000} = y$

Sketch the graphs of the following:

35. $y = \log_3 x$

36. $y = \log_{10} x$

5.2 LAWS OF LOGARITHMS

Goals

Upon completion of this section you should be able to:

1. State the three laws of logarithms.
2. Apply these laws to evaluate simple logarithmic expressions.

* *

From the definition of a logarithm we observe that *a logarithm is an exponent.* We then should expect that logarithms would have the properties of exponents and would follow the operational laws of exponents.

There are three laws of logarithms which concern us. They will allow us to multiply, divide, and find powers by logarithms.

First Law of Logarithms

The logarithm of the product of two numbers is the sum of the logarithms of the two numbers. In symbols

$$\log_b (xy) = \log_b x + \log_b y$$

Second Law of Logarithms

The logarithm of the quotient of two numbers (i.e., a fraction) is the logarithm of the numerator minus the logarithm of the denominator. In symbols

$$\log_b \frac{x}{y} = \log_b x - \log_b y$$

Third Law of Logarithms

The logarithm of a power of a number is the product of the exponent and the logarithm of the number. In symbols

$$\log_b (x)^n = n \log_b x$$

These three laws are consequences of the laws of exponents. We will demonstrate how the first law is derived. Let

$$\log_b x = k \quad \text{and} \quad \log_b y = n$$

Then $\log_b x = k$ may be written as $b^k = x$, and $\log_b y = n$ may be written as $b^n = y$. Then

$$xy = b^k b^n$$
$$= b^{k+n}$$

But $xy = b^{k+n}$ may be written as $\log_b xy = k + n$. And since $\log_b x = k$ and $\log_b y = n$,

$$\log_b xy = \log_b x + \log_b y$$

You may wish to construct a similar proof for the other two laws. We now look at some applications of these laws.

EXAMPLE

If $\log_b 5 = x$ and $\log_b 3 = y$, find $\log_b 15$.

$$\log_b 15 = \log_b (5)(3)$$
$$= \log_b 5 + \log_b 3$$
$$= x + y$$

EXAMPLE

If $\log_b 5 = x$ and $\log_b 3 = y$, find $\log_b \frac{3}{5}$.

$$\log_b \frac{3}{5} = \log_b 3 - \log_b 5$$
$$= y - x$$

EXAMPLE

If $\log_b 5 = x$, find $\log_b 125$.

$$\log_b 125 = \log_b (5)^3$$
$$= 3 \log_b 5$$
$$= 3x$$

Exercise 5.2.1

Use the laws of logarithms to find an expression for each of the following if $\log_b 5 = x$ and $\log_b 3 = y$:

1. $\log_b 9$
2. $\log_b \dfrac{3}{25}$
3. $\log_b 75$
4. $\log_b 45$
5. $\log_b \dfrac{5}{9}$
6. $\log_b \dfrac{5}{3}$
7. $\log_b \dfrac{9}{25}$
8. $\log_b 375$

Using the facts that $\log_{10} 1 = 0$ and $\log_{10} 10 = 1$, evaluate the following:

9. $\log_{10} 100$
10. $\log_{10} 1{,}000$
11. $\log_{10} 10{,}000$
12. $\log_{10} \dfrac{1}{10}$
13. $\log_{10} 0.01$
14. $\log_{10} 10^4$
15. $\log_{10} 10^5$
16. $\log_{10} 10^n$

Using the fact that $\log_{10} 3.56 = 0.5514$, evaluate the following:

17. $\log_{10} (3.56)(10^2)$
18. $\log_{10} (3.56)(10^{-1})$
19. $\log_{10} (3.56)(10^{-2})$
20. $\log_{10} (3.56)^2$
21. $\log_{10} (3.56)(10^{-3})$

5.3 TABLE OF LOGARITHMS

Goals

Upon completion of this section you should be able to:
1. Write a number in scientific notation.
2. Use a table to find the logarithm of a given number.

* *

As a preparation for finding the logarithms of numbers from logarithm tables we must first become familiar with *scientific notation*.

Remember that

$$10^0 = 1$$
$$10^1 = 10$$
$$10^2 = (10)(10) = 100$$
$$10^3 = (10)(10)(10) = 1{,}000$$
$$10^{-1} = \frac{1}{10} = 0.1$$
$$10^{-2} = \frac{1}{100} = 0.01$$
$$10^{-3} = \frac{1}{1{,}000} = 0.001$$

and so on.

To multiply a number by a power of 10, it is only necessary to move the decimal point the number of places corresponding to the exponent of 10. We move the decimal point *left* for *negative* exponents and *right* for *positive* exponents.

EXAMPLES

$2.5 \times 10^3 = 2{,}500$

$3.68 \times 10^1 = 36.8$

$2.61 \times 10^{-3} = 0.00261$

Exercise 5.3.1

Find the following products:

1. 2.3×10^2
2. 8.15×10^3
3. 8.2×10^{-4}
4. 6.31×10^{-1}
5. 7.19×10^5
6. 6.3×10^{-2}
7. 5.37×10^0
8. 8.14×10^4
9. 3.26×10^{-5}
10. 1.72×10^7

Definition

A number is expressed in *scientific notation* if it is written as a product of two factors, one of which is a number equal to or greater than 1 but less than 10 and the other is a power of 10.

Following are examples of numbers written in scientific notation:

$$26{,}800 = 2.68 \times 10^4$$
$$268 = 2.68 \times 10^2$$
$$0.0268 = 2.68 \times 10^{-2}$$
$$2.68 = 2.68 \times 10^0$$

Exercise 5.3.2

Write the following in scientific notation.

1. 46
2. 329
3. 5,280
4. 0.03
5. $\dfrac{126}{1{,}000}$
6. 37.5
7. 10,200
8. $\dfrac{315}{10{,}000{,}000}$
9. 6.19
10. $\dfrac{42}{100}$
11. 0.0301
12. 528,000

We now have three facts that will aid us in finding logarithms of numbers from a table.

1. Every decimal number can be written in scientific notation.
2. $\log_{10} 10^n = n$.
3. $\log_{10} xy = \log_{10} x + \log_{10} y$.

These facts should help us recognize that a logarithmic table (base 10) need only contain the logarithms of numbers from 1 to 10. For instance, if we wish to find the logarithm of 368 (that is, the power of 10 that will yield 368), we first write 368 in scientific notation.

$$368 = 3.68 \times 10^2$$

We now apply the first law of logarithms.

$$\log_{10}(3.68 \times 10^2) = \log_{10} 3.68 + \log_{10} 10^2$$

Then since we know that $\log_{10} 10^2 = 2$, we may write

$$\log_{10}(3.68 \times 10^2) = \log_{10} 3.68 + 2$$

5.3 TABLE OF LOGARITHMS

Now if we can find the value of $\log_{10} 3.68$ in the table, we will add 2 and have the $\log_{10} 368$. For this we need to be able to read the table of logarithms located at the end of this book.

The first column of the table is headed "N." In this column you will find numbers from 1 to 10. Notice that no decimals occur in the table. This is for the ease of printing a table, and you are expected to supply the decimals. The entry in column N of 36 is actually 3.6.

The other ten columns supply the second decimal or hundredths place.

To find the logarithm of 3.68, you look across the row from 3.6 until you reach the column headed 8 and find that

$$\log_{10} 3.68 = 0.5658$$

Notice that the decimal is also omitted in the logarithm columns. It is understood—since for $1 \leq N < 10$ we have $0 \leq \log_{10} N < 1$—that each entry is preceded by a decimal point. So

$$\log_{10} 368 = \log_{10} 3.68 + 2$$
$$= 0.5658 + 2$$
$$= 2.5658$$

From this point on we will omit the base number for base 10. If another base is desired, the base number must be written. Logarithms, base 10, are referred to as *common logarithms*. log 368 will imply base 10.

Remember these facts about the table.

1. Every entry in column N represents a number x such that $1 \leq x < 10$.
2. Every logarithm given is a number y such that $0 \leq y < 1$.

The entry in the table (logarithm of a number from 1 to 10) is referred to as the *mantissa*. The whole number part of the logarithm (the exponent of 10 when the number is in scientific notation) is called the *characteristic*. In the example just given, 0.5658 is the mantissa and 2 is the characteristic.

A step-by-step method of finding the logarithm of any number follows:

Step 1. Write the number in scientific notation.
Step 2. Use the table to find the mantissa.
Step 3. Add the characteristic (exponent of 10).

EXAMPLE   Find log 5,280.

$$\log 5{,}280 = \log(5.28 \times 10^3)$$
$$= \log 5.28 + 3$$
$$= 0.7226 + 3$$

This answer may be left as $0.7226 + 3$ or written as 3.7226.

EXAMPLE   Find log 0.0184.

$$\log 0.0184 = \log(1.84 \times 10^{-2})$$
$$= \log 1.84 - 2$$
$$= 0.2648 - 2$$

This answer may be left as $0.2648 - 2$ or written as -1.7352. Usually it is more convenient to leave it as $0.2648 - 2$. For those with calculators, the answer will appear as -1.7352 in the display window—possibly with more decimal places.

EXAMPLE   Find log 0.008.

$$\log 0.008 = \log(8.0 \times 10^{-3})$$
$$= \log 8.0 - 3$$
$$= 0.9031 - 3$$

EXAMPLE   Find log 93,400.

$$\log 93{,}400 = \log(9.34 \times 10^4)$$
$$= \log 9.34 + 4$$
$$= 0.9703 + 4$$

Exercise 5.3.3

Find the logarithms of the following numbers. Check your answers using a calculator having the logarithmic function.

1. 2.35   2. 6.82

5.3 TABLE OF LOGARITHMS

3. 1.93
4. 2.3
5. 7.9
6. 28.3
7. 37.1
8. 468
9. 329
10. 8,040
11. 32,700
12. 0.136
13. 0.014
14. 0.000508
15. 309,000
16. 0.0817
17. 0.00213
18. 683,000
19. 0.003
20. 596,000,000

21. In chemistry, the pH (hydrogen potential) of a solution is defined to be pH = $-\log [H^+]$ where $[H^+]$ is the concentration of hydrogen ions in moles per liter. Find the pH of a solution in which the concentration of hydrogen ions is 3.0×10^{-4} moles per liter.

5.4 ANTILOGARITHMS AND INTERPOLATION

Goals

Upon completion of this section you should be able to:

1. Find the antilogarithm of a number by using the table of logarithms.
2. Use interpolation to find the logarithms and antilogarithms of numbers.

* *

Definition

The number having a certain logarithm is called the *antilogarithm* of the given logarithm.

The problem, "Find the antilogarithm of 0.8235," is the same as the problem, "Find N if $\log N = 0.8235$."

To find the antilogarithms, we must reverse the procedure for finding logarithms. A step-by-step procedure follows:

Step 1. Separate the logarithm into mantissa and characteristic. (Remember that all mantissas are positive and between 0 and 1.)
Step 2. Use the table to find the antilogarithm of the mantissa.
Step 3. Use the characteristic as the exponent of 10 and multiply by the antilogarithm of the mantissa found in Step 2.

EXAMPLE   Find the antilogarithm of 3.4393.

$$3.4393 = 0.4393 + 3$$

The antilogarithm of 0.4393 is 2.75 (from table), therefore

$$\text{antilog } 3.4393 = 2.75 \times 10^3 \text{ (characteristic is 3)}$$
$$= 2,750$$

EXAMPLE   Find N if log $N = 1.6365$.

$$1.6365 = 0.6365 + 1$$
$$\text{antilog } 0.6365 = 4.33$$

Therefore
$$N = 4.33 \times 10^1$$
$$= 43.3$$

EXAMPLE   Find N if log $N = 0.3284 - 3$.

$$\text{antilog } 0.3284 = 2.13$$

Therefore
$$N = 2.13 \times 10^{-3}$$
$$= 0.00213$$

EXAMPLE   Find the antilogarithm of -2.2984.

Step 1 is a little more difficult in this case since the given logarithm is negative. It is necessary to rename -2.2984 so that the mantissa is positive. This can be accomplished by adding and then subtracting 3. Thus

$$-2.2984 = (-2.2984 + 3) - 3$$
$$= 0.7016 - 3$$

Then $\text{antilog } 0.7016 = 5.03$

Therefore $\text{antilog } 0.7016 - 3 = 5.03 \times 10^{-3}$
$$= 0.00503$$

5.4 ANTILOGARITHMS AND INTERPOLATION

Exercise 5.4.1

Find the antilogarithms of each of the following. Check answers using a calculator having the antilog or inverse log function.

1. 0.6117
2. 0.7033
3. 0.8698 + 2
4. 0.7810 − 3
5. 3.8797
6. 5.4843
7. −1.3468
8. −3.0958

The table we are using gives the logarithms of numbers between 1 and 10 if the numbers have no more than two decimal places (i.e., it is written correct to hundredths). If a number has more than two decimal places, we must find the mantissa by a procedure called *interpolation*.

If a number has more than three decimal places, it should be rounded off to three places, since interpolation for more than one place beyond that given in a table is not accurate enough to be meaningful.

A simple example might serve to convey the meaning and method of interpolation. Suppose we have a table of the squares of integers giving

$$5^2 = 25$$
$$6^2 = 36$$

and we wish to find the square of 5.5 by interpolation. Our reasoning would be, "since 5.5 is halfway between 5 and 6, then $(5.5)^2$ should be about halfway between 25 and 36." So we subtract 25 from 36 to get 11 and then add $\frac{1}{2}(11) = 5.5$ to 25 and get 30.5 as the square of 5.5. Since $(5.5)^2$ is actually 30.25, we can see that the process of interpolation is not totally accurate. However, interpolation is accurate enough for most situations. $(5.5)^2 = 30.5$ is within $\frac{1}{120}$ of the exact value.

Interpolation from the log tables given will be correct to four decimal places or within $\frac{1}{10,000}$.

EXAMPLE

Find log 1.236.

1.236 is between 1.23 and 1.24, each of which can be found in the table.

$$\log 1.23 = 0.0899$$
$$\log 1.24 = 0.0934$$

Since 1.236 is $\frac{6}{10}$ of the way between 1.23 and 1.24, we wish to find the number that is $\frac{6}{10}$ of the way from 0.0899 to 0.0934. Subtracting 0.0899 from 0.0934 gives 0.0035 and $\frac{6}{10}$ of 0.0035 is 0.0021. We add 0.0021 to 0.0899 and get

$$\log 1.236 = 0.0920$$

A pattern such as given below might be helpful.

$$10\left[6\begin{bmatrix}\log 1.230 = 0.0899\\ \log 1.236 = ?\\ \log 1.240 = 0.0934\end{bmatrix}x\right]35$$

Since we are dealing with proportional parts it is sufficient to ignore the decimals in interpolation. We actually have the proportion

$$\frac{6}{10} = \frac{x}{35}$$

Solving this we obtain

$$x = \frac{6}{10}(35)$$
$$= 21$$

Adding the 21 to 899 gives us the 920 which is, of course, the 0.0920 we seek.

EXAMPLE Find log 5,384.

$$\log 5{,}384 = \log 5.384 \times 10^3$$
$$= \log 5.384 + 3$$

$$10\left[4\begin{bmatrix}\log 5.380 = 0.7308\\ \log 5.384 = ?\\ \log 5.390 = 0.7316\end{bmatrix}x\right]8$$

5.4 ANTILOGARITHMS AND INTERPOLATION

$$\frac{4}{10} = \frac{x}{8}$$

$$x = \frac{4}{10}(8)$$

$$= 3.2 \text{ which rounds to } 3$$

(Since we are ignoring the decimals in this process, we will always round to the nearest whole number in this step.) Adding 3 to 7308 we obtain 7311. Therefore

$$\log 5.384 = 0.7311$$

and

$$\log 5{,}384 = 0.7311 + 3$$

$$= 3.7311$$

The need for interpolation of course is avoided when we use a calculator, which will give a higher degree of accuracy than our interpolation process.

Exercise 5.4.2

Find the logarithms of the following numbers. Check answers using a calculator.

1. 2.718
2. 8.003
3. 63.15
4. 569.8
5. 9,282
6. 0.01434
7. 0.2986
8. 367.4

9. It was mentioned that using interpolation to find the square of a number or the logarithm of a number is not totally accurate. Why is this so?

Interpolation to find the antilog uses the same basic idea.

EXAMPLE Find the antilog of 0.8714.

We look in the table for a mantissa of 0.8714 and find that it is between the two entries 0.8710 and 0.8716.

$$10 \left[\begin{array}{l} x \left[\begin{array}{l} \log 7.430 = 0.8710 \\ \log\ ?\ \ \ \ \ \ \ = 0.8714 \end{array} \right] 4 \\ \log 7.440 = 0.8716 \end{array} \right] 6$$

We solve the proportion

$$\frac{x}{10} = \frac{4}{6}$$

$$x = \frac{4}{6}(10)$$

$$= 6.7 \text{ which rounds to } 7$$

So $\qquad 0.8714 = \log 7.437$

or $\qquad$ antilog of $0.8714 = 7.437$

EXAMPLE If $\log N = 3.2135$, find N.

$$10\begin{bmatrix} x \begin{bmatrix} \log 1.630 = 0.2122 \\ \log ? = 0.2135 \\ \log 1.640 = 0.2148 \end{bmatrix} 13 \\ \end{bmatrix} 26$$

$$\frac{x}{10} = \frac{13}{26}$$

$$x = \frac{13}{26}(10)$$

$$= 5$$

So $\qquad 0.2135 = \log 1.635$

and $\qquad 3.2135 = \log 1.635 \times 10^3$

Thus $\qquad N = 1{,}635$

EXAMPLE Find N if $\log N = -3.9518$.

First we must change the form of -3.9518 to $0.0482 - 4$ by adding and subtracting 4. This gives us a positive mantissa of 0.0482. We then proceed as before.

$$10\begin{bmatrix} x \begin{bmatrix} \log 1.110 = 0.0453 \\ \log ? = 0.0482 \\ \log 1.120 = 0.0492 \end{bmatrix} 29 \\ \end{bmatrix} 39$$

$$\frac{x}{10} = \frac{29}{39}$$

5.4 ANTILOGARITHMS AND INTERPOLATION

$$x = \frac{29}{39} (10)$$
$$= 7.4 \text{ which rounds to } 7$$

Hence $\quad\quad\quad 0.0482 = \log 1.117$

and $\quad\quad\quad 0.0482 - 4 = \log 1.117 \times 10^{-4}$

So $\quad\quad\quad\quad\quad N = 0.0001117$

Exercise 5.4.3

Find N for each of the following. Check using a calculator.

1. $\log N = 0.7255$
2. $\log N = 2.9005$
3. $\log N = 0.2345 - 2$
4. $\log N = 3.6200$
5. $\log N = 4.0362$
6. $\log N = 0.5293 - 3$
7. $\log N = -2.6807$
8. $\log N = -0.7921$

5.5 MULTIPLICATION AND DIVISION USING LOGARITHMS

Goals

Upon completion of this section you should be able to:

1. Use logarithms to multiply two numbers.
2. Use logarithms to divide two numbers.
3. Use logarithms to solve a problem involving a combination of products and quotients.

* *

We are now prepared to perform numerical computations using logarithms, and this section will illustrate the procedure. It is expected, however, that in actual practice multiplication and division would be done using a calculator.

Two important properties will be used in much of our work with logarithms.

1. If $M = N$ (both M and N being positive), then $\log_b M = \log_b N$.
2. If $\log_b M = \log_b N$, then $M = N$.

EXAMPLE Evaluate $(5.280)(361.0)$ using logarithms. Since the log of a product is the sum of the logs (the first law of logarithms), we may write

$$\begin{aligned}
\log (5.280)(361.0) &= \log 5.28 + \log 361 = \log 5.28 + \log (3.61 \times 10^2) \\
&= 0.7226 + 2.5575 \\
&= 3.2801 \\
&= 0.2801 + 3 \\
&= \log (1.906 \times 10^3) \text{ (by interpolation)} \\
&= \log 1{,}906
\end{aligned}$$

Hence $(5.280)(361.0) = 1{,}906$.

EXAMPLE Evaluate $\dfrac{0.0065}{0.0732}$ using logarithms. Using the second law of logarithms, we write

$$\begin{aligned}
\log \frac{0.0065}{0.0732} &= \log 0.0065 - \log 0.0732 \\
&= \log (6.5 \times 10^{-3}) - \log (7.32 \times 10^{-2}) \\
&= (0.8129 - 3) - (0.8645 - 2) \\
&= 0.0516 - 1 \\
&= 0.9484 - 2 \text{ (by adding and subtracting 1)} \\
&= \log (8.88 \times 10^{-2}) \text{ (no interpolation is necessary)} \\
&= \log 0.0888
\end{aligned}$$

Therefore $\dfrac{0.0065}{0.0732} = 0.0888$

EXAMPLE Evaluate $\dfrac{(7.326)(0.0731)}{(6.28)(3.14)}$ using logarithms. Using the first two laws of logarithms, we write

$$\begin{aligned}
\log \frac{(7.326)(0.0731)}{(6.28)(3.14)} &= (\log 7.326 + \log 0.0731) - (\log 6.28 + \log 3.14) \\
&= (0.8649 + 0.8639 - 2) - (0.7980 + 0.4969) \\
&= (1.7288 - 2) - (1.2949)
\end{aligned}$$

5.5 MULTIPLICATION AND DIVISION USING LOGARITHMS

$$= 0.4339 - 2$$
$$= \log 2.716 \times 10^{-2}$$
$$= \log 0.02716$$

Thus $\quad\dfrac{(7.326)(0.0731)}{(6.28)(3.14)} = 0.02716$

Exercise 5.5.1

Evaluate the following using logarithms. Check without logarithms by using a calculator.

1. $(2.3)(5.2)$
2. $(5.19)(9.01)$
3. $(23)(58)$
4. $(593)(26.8)$
5. $\dfrac{3.57}{2.11}$
6. $\dfrac{29.3}{2.69}$
7. $\dfrac{0.0154}{0.0023}$
8. $\dfrac{0.254}{34.6}$
9. $\dfrac{(245)(30.1)}{(58.6)(24)}$
10. $\dfrac{(0.001593)(23.8)}{(0.014)(25.5)}$

5.6 FINDING POWERS AND ROOTS USING LOGARITHMS

Goals

Upon completion of this section you should be able to:

1. Use logarithms to find the power of a number such as $(15.3)^5$.
2. Use logarithms to find the roots of numbers such as $\sqrt{0.0951}$.

* *

The third law of logarithms states
$$\log (x)^n = n \log x$$
This law is used where n is any rational number.

LOGARITHMS

EXAMPLE Evaluate $(0.0432)^8$ using logarithms.

$$\begin{aligned}\log (0.0432)^8 &= 8 \log 0.0432 \\ &= 8 (0.6355 - 2) \\ &= 5.0840 - 16 \\ &= 0.0840 - 11 \\ &= \log (1.213 \times 10^{-11}) \\ &= \log 0.00000000001213\end{aligned}$$

Therefore $\qquad (0.0432)^8 = 0.00000000001213$

EXAMPLE Evaluate $\sqrt[5]{125}$ using logarithms.

$$\begin{aligned}\log \sqrt[5]{125} &= \log (125)^{\frac{1}{5}} \\ &= \frac{1}{5} \log 125 \\ &= \frac{1}{5} (2.0969) \\ &= 0.4194 \\ &= \log 2.627\end{aligned}$$

Therefore $\qquad \sqrt[5]{125} = 2.627$

EXAMPLE Evaluate $\sqrt[3]{0.0357}$ using logarithms.

$$\begin{aligned}\log \sqrt[3]{0.0357} &= \log (0.0357)^{\frac{1}{3}} \\ &= \frac{1}{3} \log 0.0357 \\ &= \frac{1}{3} (0.5527 - 2)\end{aligned}$$

Since $0.5527 - 2$ divided by 3 would give a fractional characteristic, it is convenient to rewrite the expression as $1.5527 - 3$. Then

$$\begin{aligned}\log \sqrt[3]{0.0357} &= \frac{1}{3} (1.5527 - 3) \\ &= 0.5176 - 1 \\ &= \log (3.293 \times 10^{-1})\end{aligned}$$

5.6 FINDING POWERS AND ROOTS USING LOGARITHMS

$$= \log 0.3293$$

Thus $\sqrt[3]{0.0357} = 0.3293$

Exercise 5.6.1

Evaluate the following using logarithms. Check answers using a calculator with a power function (remember you may convert roots to fractional powers).

1. $(2.15)^3$
2. $(16.3)^2$
3. $(0.21)^4$
4. $(15.3)^5$
5. $(0.053)^7$
6. $\sqrt{55}$
7. $\sqrt{256}$
8. $\sqrt[3]{2,230}$
9. $\sqrt{0.0951}$
10. $\sqrt[5]{0.687}$

5.7 LOGARITHMS TO DIFFERENT BASES

Goals

Upon completion of this section you should be able to:

1. State the formula for finding the logarithm of a number to any desired base.
2. Find logarithms of numbers to various bases using the base 10 table.
3. Find logarithms to base e.

* *

Sometimes it is necessary to change the base of a logarithm. This is accomplished by the following formula:

$$\log_b x = \frac{\log_a x}{\log_a b}$$

The formula is obtained as follows. Let

$$\log_b x = N$$

then

$$b^N = x$$

If we then take the log of each side to the base a, we obtain

$$\log_a b^N = \log_a x$$

or

$$N \log_a b = \log_a x$$

LOGARITHMS

Then solving for N we obtain the desired result

$$N = \frac{\log_a x}{\log_a b}$$

EXAMPLE Find $\log_5 45$.

It is not necessary to have a table of logarithms to base 5 to find this logarithm. We may use our base 10 table by applying the formula.

$$\log_5 45 = \frac{\log 45}{\log 5}$$

Find the values of log 45 and log 5 in our base 10 table we have

$$\log_5 45 = \frac{1.6532}{0.6990}$$

We are now faced with the problem of dividing 1.6532 by 0.6990.

CAUTION: Do not make the mistake of subtracting these numbers. $\frac{\log 45}{\log 5}$ bears no relationship to $\log 45 - \log 5$.

We must either perform this computation by long division or by using logarithms. We will use logarithms.

$$\log \frac{1.6532}{0.6990} = \log 1.6532 - \log 0.6990$$

$$= 0.2183 - (0.8445 - 1)$$

$$= 0.3738$$

$$= \log 2.365$$

Therefore $\qquad \dfrac{1.6532}{0.6990} = 2.365$

Hence $\qquad \log_5 45 = 2.365$

Note that the original problem in the example, find $\log_5 45$, could have been stated "Find x if $5^x = 45$," since

$$\log_5 45 = 2.365$$

in exponential form is $\qquad 5^{2.365} = 45$

The *natural number* "e" (an irrational number) is often used in higher mathematics as a base for logarithmic tables. "e" correct to four digits is 2.718.

5.7 LOGARITHMS TO DIFFERENT BASES

In Problem 1 of Exercise 5.4.2 you were asked to find log 2.718. The value was 0.4343. Thus we know that

$$\log e = 0.4343$$

With this value we can change from base 10 (common logarithms) to base e (natural logarithms). Since natural logarithms are used so much in mathematics, we use special notation to represent them. The expression ln x means "the logarithm of x to the base e." Thus ln x = $\log_e x$.

EXAMPLE

Find ln 25.

$$\ln 25 = \frac{\log 25}{\log e}$$

$$= \frac{1.3979}{0.4343}$$

Again using logarithms

$$\log \frac{1.3979}{0.4343} = \log 1.3979 - \log 0.4343$$

$$= 0.1454 - (0.6378 - 1)$$

$$= 0.5076$$

$$= \log 3.218$$

Therefore $\quad \dfrac{1.3979}{0.4343} = 3.218$

Hence $\quad \ln 25 = 3.218$

Exercise 5.7.1

Find the following logarithms:

1. $\log_2 29$
2. $\log_5 35$
3. $\log_9 613$
4. $\log_3 0.035$
5. $\log_{24} 793$
6. ln 346
7. ln 50.1
8. ln 0.0082

5.8 EXPONENTIAL AND LOGARITHMIC EQUATIONS

Goals

Upon completion of this section you should be able to:

1. Solve logarithmic equations such as $\log (x + 1) = \log x + \log 3$.
2. Solve exponential equations such as $2^{x+1} = 3^{2x+5}$.

* *

In certain equations the variable appears as an exponent or in a logarithmic expression. Many such equations may be solved as in the following examples.

EXAMPLE

Solve for x: $\log (x + 1) = \log x + \log 3$.

In general, when the variable appears in a logarithmic expression, we can find a solution by taking the antilog of both sides of the equation. This requires that we apply the laws of logarithms in reverse.

$$\log (x + 1) = \log x + \log 3$$
$$\log (x + 1) = \log 3x$$
$$x + 1 = 3x$$
$$x = \frac{1}{2}$$

The solutions for equations such as this should be checked by substituting the value back into the original equation.

$$\log (x + 1) = \log \left(\frac{1}{2} + 1\right) = \log \frac{3}{2}$$

and

$$\log x + \log 3 = \log 3x = \log (3)\left(\frac{1}{2}\right) = \log \frac{3}{2}$$

Therefore $\frac{1}{2}$ is a solution to the equation.

EXAMPLE

Solve for x.

$$\log (2x - 1) = \log (x + 2) + \log (x - 2)$$
$$\log (2x - 1) = \log (x + 2)(x - 2)$$

$$2x - 1 = (x + 2)(x - 2)$$
$$2x - 1 = x^2 - 4$$
$$x^2 - 2x - 3 = 0$$
$$(x - 3)(x + 1) = 0$$
$$x = 3 \quad \text{or} \quad x = -1$$

Checking our answers we see that 3 is a solution, but if we substitute $x = -1$ in the last term of the equation $\log(x - 2)$, we obtain $\log(-3)$. But we cannot find the logarithm of a negative number. Thus -1 is not a solution of the original equation. Therefore the solution is 3.

EXAMPLE

Solve for x: $3^x = 5$.

In this equation the variable occurs in the exponent. In such cases it is usually necessary to take the log of both sides of the equation.

$$3^x = 5$$
$$\log 3^x = \log 5$$
$$x \log 3 = \log 5$$
$$x = \frac{\log 5}{\log 3}$$

We will leave the solution in terms of logarithms. To find a numerical answer would require only that we use the table to find log 5 and log 3 and then divide.

EXAMPLE

Solve for x: $2^{x+1} = 8$.

Taking the log of each side we obtain

$$\log 2^{x+1} = \log 8$$
$$\log 2^{x+1} = \log 2^3$$
$$(x + 1) \log 2 = 3 \log 2$$

Dividing each side by log 2 yields

$$x + 1 = 3$$
$$x = 2$$

EXAMPLE

Solve for x: $3^{x-4} = 2^{x+1}$.

Taking the log of each side we obtain

$$\log 3^{x-4} = \log 2^{x+1}$$
$$(x - 4) \log 3 = (x + 1) \log 2$$
$$x \log 3 - 4 \log 3 = x \log 2 + \log 2$$
$$x \log 3 - x \log 2 = \log 2 + 4 \log 3$$
$$x (\log 3 - \log 2) = \log 2 + 4 \log 3$$
$$x = \frac{\log 2 + 4 \log 3}{\log 3 - \log 2}$$

Applications of logarithms are found in areas of business and science. The following example is an illustration of the use of logarithms in biology.

EXAMPLE

The number N of bacteria present in a certain culture at the end of t hours is given by $N = N_0 \times 10^{0.01t}$, where N_0 is the original number of bacteria. How many hours will it take for the number of bacteria to double?

When the number of bacteria doubles, the value of N will be twice that of N_0. Thus, the formula will read

$$2N_0 = N_0 \times 10^{0.01t}$$

Dividing each side by N_0, we obtain

$$2 = 10^{0.01t}$$

We now take the logarithm of each side, obtaining

$$\log 2 = \log 10^{0.01t}$$

Using the third law of logarithms, we have

$$\log 2 = 0.01t \log 10$$

Since $\log 10 = 1$, we may write

$$\log 2 = 0.01t$$

or

$$t = \frac{\log 2}{0.01}$$

Evaluating log 2 yields

5.8 EXPONENTIAL AND LOGARITHMIC EQUATIONS

$$t = \frac{0.3010}{0.01} = 30.1 \text{ hours}$$

Exercise 5.8.1

Solve the following equations:

1. $\log(x + 2) = \log x + \log 2$
2. $\log(5x - 3) = \log(2x + 1) + \log 2$
3. $\log(3x + 1) = \log(x + 2) + \log 4$
4. $\log(x + 3) = \log(2x - 1) - \log 3$
5. $\log(x + 5) - \log 2 = \log(5x + 2)$
6. $2 \log x = \log 9$
7. $2 \log(x + 3) = \log(x + 1) + \log(x + 4)$
8. $2 \log(x + 1) = \log(5x + 11)$
9. $\log(x - 1) + \log(x - 2) = \log 2 + \log(2x - 5)$
10. $\log(x + 6) = \log(x + 20) - \log(x + 2)$
11. $3^{x-2} = 9$
12. $5^{2x-1} = 125$
13. $2^x = 7$
14. $6^{x-1} = 11$
15. $2^{3x+4} = 9$
16. $7^{2x-3} = 1$
17. $2^{x+1} = 3^{2x+5}$
18. $5^{2x-1} = 11^{x+5}$
19. $3^{x-2} = 15^{2x-1}$
20. $\log(x + 3) + \log 5 = 2$
21. $\log 3 = 3 - \log(x - 2)$
22. $\log x + \log(x + 15) = 2$

23. The amount A present in an account that pays r percent annual interest for n years is given by the formula $A = p(1 + r)^n$ where p is the principal invested. If the annual interest rate is 8%, how many years will it take an investment to double in value?

24. The number of bacteria N in a certain culture at the end of t hours is given by $N = 1,000 \times 10^{0.1t}$. Find the number of bacteria present at the end of 3 hours.

25. The amount A of radioactive carbon 14 remaining at any time t is given by the formula $A = A_0(2)^{-\frac{t}{5750}}$ where A_0 is the amount that was originally present. An animal's tooth was tested and found to have lost 60% of its carbon 14 content. How old is the animal?

26. The intensity I (in lumens) of a beam of light passing through a substance is given by the formula $I = I_0 e^{-at}$ where I_0 is the intensity of the light before passing through the substance, a is the absorption coefficient of the substance, and t is the thickness (in centimeters) of the substance. If the initial intensity is 1,000 lumens and the absorption coefficient of the substance is 0.2, what thickness is necessary to reduce the illumination to 600 lumens?

CHAPTER REVIEW

Solve for the variable.

1. $\log_4 x = 3$
2. $\log_2 8 = y$
3. $\log_b 81 = 2$
4. $\log_2 x = -3$
5. $\log_3 \dfrac{1}{27} = y$

If $\log_b 2 = x$ and $\log_b 3 = y$, find the following:

6. $\log_b 6$
7. $\log_b 8$
8. $\log_b \dfrac{2}{9}$
9. $\log_b 72$
10. $\log_b \dfrac{4}{27}$

Find the logarithms of the following numbers:

11. 8.04
12. 38.6
13. 49,000
14. 0.0062
15. 0.000156

Find N for each of the following:

16. $\log N = 0.6405$
17. $\log N = 2.9581$
18. $\log 173.4 = N$
19. $\log 0.4507 = N$
20. $\log N = 0.8683 - 2$

Evaluate using logarithms.

21. $(2.7)(8.1)$
22. $(25)(470)$
23. $(0.013)(6.24)$
24. $\dfrac{381}{29.3}$
25. $\dfrac{0.034}{0.263}$

Evaluate using logarithms.

26. $(38.7)^3$
27. $(0.052)^3$
28. $\sqrt{529}$
29. $\sqrt[3]{71.4}$
30. $\sqrt[5]{0.0342}$

Find the following logarithms:

31. $\log_4 104$
32. $\log_2 33.6$
33. $\log_5 2.27$
34. $\ln 912$
35. $\ln 0.826$

Solve the following:

36. $\log(3x - 1) = \log 2 + \log(x + 3)$
37. $\log(x - 6) = \log(x + 4) - \log 3$
38. $\log(x + 1) + \log(x + 2) = \log(7x + 23)$
39. $3^{x+1} = 7$
40. $2^{x-1} = 5^{2x+1}$
41. "The Law of Natural Growth" is given by the formula $A = A_0 e^{rn}$ where A is the resulting amount, A_0 is the original amount, r is the rate of growth, and n is the time. If a city has a steady growth rate of 5%, in how many years will the population double?

PRACTICE TEST

1. Write in logarithmic form $3^2 = 9$.
2. Write in exponential form $\log_2 \frac{1}{4} = -2$.
3. Find a in each of the following:
    a. $\log_4 16 = a$
    b. $\log_3 a = 2$
    c. $\log_a 27 = 3$
    d. $\log_2 \frac{1}{16} = a$
    e. $\log 10^5 = a$
4. If $\log_b 2 = x$ and $\log_b 3 = y$, find the following:
    a. $\log_b 9$
    b. $\log_b 12$
5. Find the following logarithms:
    a. $\log 24.7$
    b. $\log 0.000156$
    c. $\log 7.143$
    d. $\log_4 316$
6. Find N.
    a. $\log N = 0.8331$
    b. $\log N = 3.4742$
    c. $\log N = 0.7364 - 2$

7. Evaluate using logarithms.

   a. $(23.6)^3$ b. $\sqrt{547}$

8. Solve for x.

   a. $\log(x + 2) = \log 5 + \log x$ b. $3^{x-1} = 81$

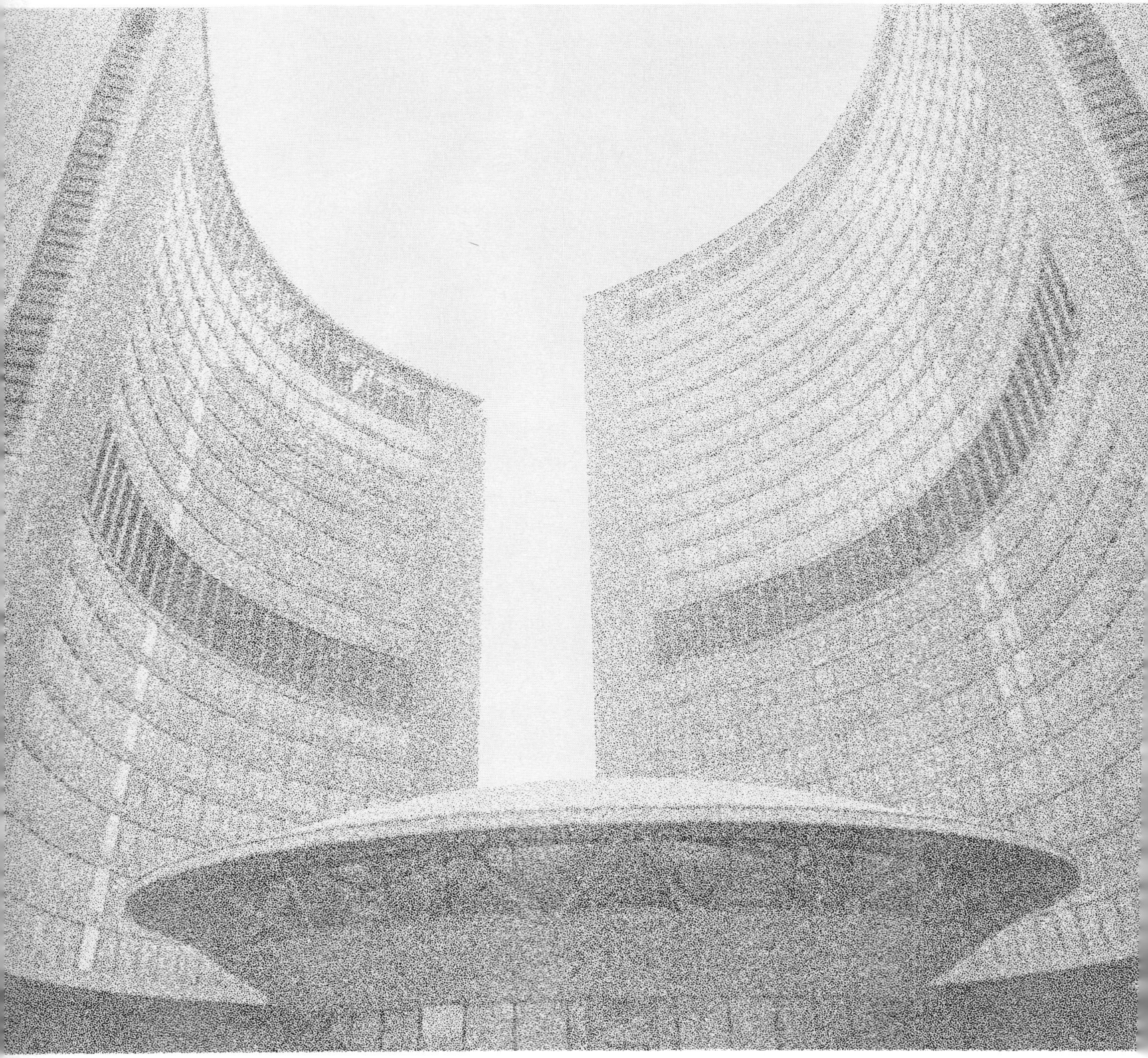

Polynomials

6

6

In Sec. 1.3 we defined a monomial, and then defined a polynomial as a sum of monomials. The student should probably review that definition. Since that section we have multiplied, factored, and simplified polynomials, and we have solved some polynomial equations. In this chapter we wish to develop some of the theory of polynomials and their associated polynomial equations.

6.1 PROPERTIES OF POLYNOMIALS

Goals

Upon completion of this section you should be able to:

1. Determine the degree of a polynomial.
2. Determine if a given value is a zero of a given polynomial.
3. Use the properties of an integral domain in working with polynomials.

* *

The functional notation discussed in Chapter 3 can be useful in the study of polynomials. In general we will denote a polynomial in x as $P(x)$.

The student should convince himself/herself that every polynomial P does define a function. In other words, for a given polynomial P in x, does each value x_0 yield one and only one value $P(x_0)$?

A general polynomial function is usually written as

$$P(x) = a_0 x^n + a_1 x^{n-1} + a_2 x^{n-2} + \cdots + a_{n-1} x + a_n$$

The coefficients a_i and the variable x may be either real or complex.

1. $P(x)$ is a *polynomial over the real number field* if the coefficients are real numbers (that is, $a_i \in R$).
2. $P(x)$ is a *polynomial in the real variable* x if the variable x can take on only real values (that is, $x \in R$).

3. $P(x)$ is a *real polynomial* if both the coefficients and the variable are real numbers (that is, $a_i \in R$ and $x \in R$).

We may state similar definitions involving complex numbers if we substitute the word "complex" for the word "real" in each of the three statements.

Definition The *degree* of a polynomial in x is the highest exponent of x that appears in the polynomial.

For instance,
$$P(x) = 2x^3 + 5x^2 - 3 \text{ is of degree } 3$$
$$P(x) = 6x^5 - 2x^3 + x - 1 \text{ is of degree } 5$$
$$P(x) = 9x - 2 \text{ is of degree } 1$$
$$P(x) = 5 \text{ is of degree } 0$$

It is generally agreed that no degree is assigned to $P(x) = 0$.

Definition If for a polynomial P in x, $P(a) = 0$, then a is a *zero* of the polynomial.

For instance, if
$$P(x) = x^2 + 5x + 6$$
then since
$$P(-2) = (-2)^2 + 5(-2) + 6 = 0$$
we conclude that -2 is a zero of the polynomial.

In solving linear and quadratic equations, we already have the tools for finding the zeros of first- and second-degree polynomials. Later in this chapter we will discuss zeros of higher-degree polynomials.

Because we have already added and multiplied polynomials, we should address ourselves to the question of what properties apply over the entire set of polynomials.

All but one of the field properties which were given for the set of real numbers hold true for the set of polynomials. Let us refer back to Sec. 1.1 and examine each of these properties as they relate to polynomials.

1. The set of polynomials is *closed* under the operation of addition since the sum of two polynomials is also a polynomial.
2. Addition of polynomials is *commutative* since $P(x) + Q(x) = Q(x) + P(x)$ for all polynomials.
3. Addition of polynomials is *associative*. $[P(x) + Q(x)] + R(x) = P(x) + [Q(x) + R(x)]$ for all polynomials.

4. There is a polynomial which is the *additive identity* for the set of all polynomials. That is, there is some polynomial $J(x)$, such that $P(x) + J(x) = J(x) + P(x) = P(x)$ for all polynomials $P(x)$. You are asked to find this polynomial in the exercises.
5. Each polynomial has an *additive inverse* [denoted by $-P(x)$] such that $P(x) + [-P(x)] = J(x)$. An example of this is asked for in the exercises.
6. The set of polynomials is *closed* under multiplication since the product of two polynomials is a polynomial.
7. Multiplication of polynomials is *commutative*. $P(x) \cdot Q(x) = Q(x) \cdot P(x)$ for all polynomials.
8. Multiplication of polynomials is *associative*. $[P(x) \cdot Q(x)] \cdot R(x) = P(x) \cdot [Q(x) \cdot R(x)]$ for all polynomials.
9. There is a polynomial which is the *multiplicative identity*. There exists some polynomial $K(x)$ such that $P(x) \cdot K(x) = K(x) \cdot P(x) = P(x)$ for all polynomials $P(x)$. We find that this polynomial is $K(x) = 1$.
10. The only field axiom that does not hold for polynomials is the axiom involving a multiplicative inverse for each nonzero element. (In order to show that an axiom does not hold, it is only necessary to show one counterexample.)

Problem Does axiom 10 from Sec. 1.1 hold true for the set of polynomials?

In other words, does each polynomial have a multiplicative inverse, or for $P(x)$ is there a polynomial $Q(x)$ such that

$$P(x) \cdot Q(x) = 1$$

To show a counterexample, we can use the polynomial $5x^2$. If $5x^2$ has a multiplicative inverse $Q(x)$, then

$$5x^2 \cdot Q(x) = 1$$

and

$$Q(x) = \frac{1}{5x^2}$$

but $Q(x) = \frac{1}{5x^2}$ is not a polynomial. Therefore $5x^2$ does not have a multiplicative inverse in the set of polynomials, and the set of polynomials does *not* form a field.

11. Multiplication is *distributive* over addition in the set of polynomials. $P(x) [Q(x) + R(x)] = P(x) \cdot Q(x) + P(x) \cdot R(x)$ for all polynomials. An example is asked for in the exercises.

A set of elements that satisfies all of the field properties except that of the multiplicative inverse is called an *integral domain.* Because the set of polynomials does satisfy all field axioms except for the fact that not every polynomial has a multiplicative inverse that is a polynomial, we say that the set of polynomials forms an integral domain.

Even though we do not have a field, we may still use those axioms which do hold true for our manipulations with polynomials.

Exercise 6.1.1

1. If $P(x) = x^4 + x^3 - 13x^2 - x + 12$, find

   a. $P(2)$ b. $P(-1)$ c. $P(0)$ d. $P(3)$ e. $P(-4)$ f. $P(1)$

2. Did you find any values for x in Problem 1 that are zeros of $P(x)$? If so, which values are zeros? How many zeros did you find?

3. Is -2 a zero of $P(x) = x^5 + 2x^3 - 27x^2 - 54$?

4. Is 0 a zero of $P(x) = 3x^6 - 144x^5 + 29x^4 - 78x^2 - 1$?

5. What is the additive identity for the set of polynomials?

6. Give the additive inverse of $P(x) = x^3 + 3x^2 - x + 5$.

7. Do *any* polynomials have multiplicative inverses? Explain.

8. Let $P(x) = x^2 + 1$, $Q(x) = x^3 + 2x^2 - 1$, and $R(x) = 4x^3 + x + 4$. Show that $P(x) [Q(x) + R(x)] = P(x) \cdot Q(x) + P(x) \cdot R(x)$.

9. Let $P(x) = x^3 + 5x^2 + 2x - 8$ and $Q(x) = 3x^6 + 4x^5 - 5x^3 + 1$. State the degree of

   a. $P(x)$ b. $Q(x)$ c. $P(x) + Q(x)$ d. $P(x) \cdot Q(x)$

10. Two polynomials in x are said to be equal if and only if they are of the same degree and their respective coefficients of terms of the same degree are equal. If $P(x) = 5x^2 - 2x + 1$ and $Q(x) = (a + 4)x^2 + (b - 9)x + 1$, find a and b so that $P(x) = Q(x)$.

6.2 DIVISION OF POLYNOMIALS

Goals

Upon completion of this section you should be able to:

1. Use long division to divide polynomials.
2. Find the remainder by using the remainder theorem.
3. Determine factors of a polynomial by using the factor theorem.

* *

In previous chapters we have added, subtracted, and multiplied polynomials. In this section we wish to discuss the other fundamental operation—division.

From arithmetic we are familiar with a process called "long division." For instance, to divide 30 by 8 using long division we write

$$8 \overline{)\,30\,} \quad \begin{array}{c} 3 \\ \underline{24} \\ 6 \end{array}$$

and say that 30 divided by 8 gives a quotient of 3 and a remainder of 6. We can express this as

$$30 = 8(3) + 6$$

In fact, for integers a and b where $b < a$ we can write

$$a = bq + r$$

where q is the quotient and r is the remainder such that r is less than b. This is the *division algorithm for integers*.

EXAMPLE

Find the quotient and remainder when 2,876 is divided by 27.

$$27 \overline{)\,2{,}876\,} \quad \begin{array}{c} 106 \\ \underline{27} \\ 176 \\ \underline{162} \\ 14 \end{array}$$

so $\qquad\qquad 2{,}876 = 27(106) + 14$

and $\qquad\qquad q = 106,\, r = 14$

A similar process can be used to divide one polynomial by another.

EXAMPLE

Divide $6x^3 - 17x^2 - x + 15$ by $(2x - 5)$.

Step 1 Make sure that both polynomials are arranged in descending powers of x. Supply zero coefficients for any missing terms and arrange as follows:

$$2x - 5 \,\overline{)\,6x^3 - 17x^2 - x + 15\,}$$

192   POLYNOMIALS

Step 2 To obtain the first term of the quotient, divide the first term of the dividend by the first term of the divisor. In this case, $6x^3 \div 2x$ gives $3x^2$. Write as follows:

$$\begin{array}{r} 3x^2 \\ 2x-5 \overline{\smash{\big)}\, 6x^3 - 17x^2 - x + 15} \end{array}$$

Step 3 Multiply the *entire* divisor by the term obtained in Step 2 and subtract the result from the dividend as follows:

$$\begin{array}{r} 3x^2 \\ 2x-5 \overline{\smash{\big)}\, 6x^3 - 17x^2 - x + 15} \\ \underline{6x^3 - 15x^2 } \\ -\ 2x^2 - x + 15 \end{array}$$

Step 4 Divide the first term of the remainder by the first term of the divisor to obtain the next term of the quotient. Then multiply the entire divisor by the term thus obtained and subtract again as follows:

$$\begin{array}{r} 3x^2 - x \\ 2x-5 \overline{\smash{\big)}\, 6x^3 - 17x^2 - x + 15} \\ \underline{6x^3 - 15x^2 } \\ -\ 2x^2 - x + 15 \\ \underline{-\ 2x^2 + 5x } \\ -\ 6x + 15 \end{array}$$

This process is repeated until the remainder is zero or a polynomial of degree *less* than the divisor.

$$\begin{array}{r} 3x^2 - x - 3 \\ 2x-5 \overline{\smash{\big)}\, 6x^3 - 17x^2 - x + 15} \\ \underline{6x^3 - 15x^2 } \\ -\ 2x^2 - x + 15 \\ \underline{-\ 2x^2 + 5x } \\ -\ 6x + 15 \\ \underline{-\ 6x + 15} \\ 0 \end{array}$$

Therefore our example shows that

$$(6x^3 - 17x^2 - x + 15) \div (2x - 5) = 3x^2 - x - 3$$

or

$$6x^3 - 17x^2 - x + 15 = (2x - 5)(3x^2 - x - 3) + 0$$

6.2 DIVISION OF POLYNOMIALS

EXAMPLE

Divide $5x^3 - 24x^2 + 7$ by $(5x + 1)$. Notice that the dividend is missing an x term. This is because the coefficient of that term is zero. We must remember to enter the x term as $0x$ as follows:

$$
\begin{array}{r}
x^2 - 5x + 1 \\
5x + 1 \overline{\smash{)}5x^3 - 24x^2 + 0x + 7} \\
\underline{5x^3 + x^2 } \\
-25x^2 + 0x + 7 \\
\underline{-25x^2 - 5x } \\
5x + 7 \\
\underline{5x + 1} \\
6
\end{array}
$$

so $\quad 5x^3 - 24x^2 + 7 = (5x + 1)(x^2 - 5x + 1) + 6$

These examples have illustrated the *division algorithm for polynomials*. We will state it here without proof.

Theorem

If $f(x)$ and $g(x)$ are polynomials and if $g(x) \neq 0$, then there exist unique polynomials $q(x)$ and $r(x)$ such that $f(x) = g(x) \cdot q(x) + r(x)$ where either $r(x) = 0$ or $r(x)$ has degree less than the degree of $g(x)$.

In this theorem $q(x)$ is the quotient and $r(x)$ is the remainder.

A valuable special case occurs when $f(x)$ is divided by a first-degree polynomial such as $(x - c)$. The theorem informs us that $r(x)$ would have degree zero. Why? A polynomial of degree zero is, of course, a real number.

We can write

$$f(x) = (x - c)\, q(x) + d$$

where d is a real number. If we now substitute $x = c$ in the expression, we obtain

$$f(c) = (c - c)\, q(c) + d$$

which reduces to $\quad f(c) = d$

giving us $\quad f(x) = (x - c)\, q(x) + f(c)$

We have just provided proof for the following theorem.

Remainder Theorem

If a polynomial $f(x)$ is divided by $(x - c)$, the remainder is $f(c)$.

EXAMPLE

Find the remainder if $3x^2 - 2x + 7$ is divided by $(x - 3)$: (*a*) using long division and (*b*) using the remainder theorem.

a. By long division

$$\begin{array}{r} 3x + 7 \\ x - 3 \overline{\smash{\big)}\, 3x^2 - 2x + 7} \\ \underline{3x^2 - 9x} \\ 7x + 7 \\ \underline{7x - 21} \\ 28 \end{array}$$

Thus the remainder is 28.

b. By the remainder theorem

$$f(x) = 3x^2 - 2x + 7$$
$$f(3) = 3(3)^2 - 2(3) + 7$$
$$= 28$$

The remainder is 28.

EXAMPLE

Find the remainder, using the remainder theorem, if $5x^2 - 7x + 12$ is divided by $(x + 4)$. Here we must note that the theorem states that we divide by $(x - c)$. To use the theorem, we must express $(x + 4)$ in terms of $(x - c)$.

$$(x + 4) = [x - (-4)]$$

so that $c = -4$.

$$f(x) = 5x^2 - 7x + 12$$
$$f(-4) = 120$$

Thus if $5x^2 - 7x + 12$ is divided by $(x + 4)$, the remainder is 120.

Another theorem closely related to the remainder theorem is the factor theorem.

Factor Theorem

A polynomial $f(x)$ has a factor $(x - c)$ if and only if $f(c) = 0$.

The words "if and only if" in this theorem mean that the statement and its converse are both true. That is, the statements:

and
$$\text{If } f(c) = 0, \text{ then } f(x) \text{ has a factor } (x - c)$$
$$\text{If } f(x) \text{ has a factor } (x - c), \text{ then } f(c) = 0$$
are both true. We must, therefore, prove both statements.

Proof If $f(c) = 0$, then by the division algorithm and the remainder theorem

$$f(x) = (x - c) q(x) + 0$$

or

$$f(x) = (x - c) q(x)$$

and therefore $(x - c)$ is a factor of $f(x)$.

Conversely, if $(x - c)$ is a factor of $f(x)$, then the remainder $f(c)$ is zero since being a factor implies division with a remainder of zero.

EXAMPLE

Determine if $(x - 2)$ is a factor of $x^3 - 6x^2 + 11x - 6$ by the factor theorem.

$$f(x) = x^3 - 6x^2 + 11x - 6$$
$$f(2) = (2)^3 - 6(2)^2 + 11(2) - 6$$
$$= 0$$

Hence $(x - 2)$ is a factor of $x^3 - 6x^2 + 11x - 6$.

EXAMPLE

Is $(x + 1)$ a factor of $x^3 - 6x^2 + 11x - 6$?

$$f(x) = x^3 - 6x^2 + 11x - 6$$
$$f(-1) = (-1)^3 - 6(-1)^2 + 11(-1) - 6$$
$$= -24$$

Since $f(-1) \neq 0$, $(x + 1)$ is not a factor of $x^3 - 6x^2 + 11x - 6$.

EXAMPLE

Check to see if both $(x + 3)$ and $\left(x - \frac{1}{2}\right)$ are factors of the polynomial $2x^3 + 3x^2 - 8x + 3$ using the factor theorem.

$$f(x) = 2x^3 + 3x^2 - 8x + 3$$
$$f(-3) = 2(-3)^3 + 3(-3)^2 - 8(-3) + 3$$
$$= 0$$

Hence $(x + 3)$ is a factor. Next, determine if $\left(x - \frac{1}{2}\right)$ is a factor; check to see if $f\left(\frac{1}{2}\right)$ is equal to zero.

$$f\left(\frac{1}{2}\right) = 2\left(\frac{1}{2}\right)^3 + 3\left(\frac{1}{2}\right)^2 - 8\left(\frac{1}{2}\right) + 3$$
$$= 0$$

Thus both $(x + 3)$ and $\left(x - \frac{1}{2}\right)$ are factors of the polynomial.

Exercise 6.2.1

1. Find the quotient and remainder for the following divisions:
   a. $(3x^3 - x^2 + 4x - 2) \div (x + 1)$
   b. $(2x^4 + 13x^3 + 15x^2 - 4x - 9) \div (2x + 3)$
   c. $(3x^5 - 4x^3 + x + 1) \div (x^2 - 2)$
   d. $(x^5 - 13x^3 + 11x^2 + 15) \div (x + 4)$
   e. $(x^5 - x^4 + 2x^3 - x + 10) \div (x^2 + 1)$
   f. $(6x^4 - 51x^2 - 27) \div (x - 3)$
   g. $(2x^4 + 5x^3 - 11x^2 - 20x + 12) \div (x^2 + x - 6)$

2. In each of the following divisions find the remainder by (1) long division and (2) the remainder theorem.
   a. $(2x^3 - x^2 + 1) \div (x - 1)$ b. $(5x^2 + x - 4) \div (x - 4)$
   c. $(x^5 + 3x^2 - x) \div (x + 1)$ d. $(3x^4 + 5x^3 + 2) \div (x + 3)$
   e. $(2x^4 + x^3 - 16x^2 + 18) \div (x + 2)$ f. $(x^5 - 32) \div (x - 2)$

3. Is $(x + 4)$ a factor of $3x^3 + 5x^2 - 2x - 1$?
4. Is $(x + 4)$ a factor of $5x^3 + 16x^2 - 18x - 8$?
5. Is $(x - 5)$ a factor of $3x^3 - 17x^2 + 11x - 5$?
6. Is $(x + 1)$ a factor of $5x^4 + 10x^3 - 15x^2 + 24x + 16$?
7. Is $\left(x - \frac{1}{2}\right)$ a factor of $4x^4 - 8x^3 + 3x^2 + 8x - 4$?
8. If $P(x) = x^3 - x^2 - 16x - 12$, is $(x + 3)$ a factor of $P(x)$? Can you determine a zero for this polynomial?

6.3 SYNTHETIC DIVISION

Goals

Upon completion of this section you should be able to:

1. Use synthetic division to divide polynomials.
2. Find the remainder using synthetic division.

* *

Whenever a process is used often in algebra, it helps to shorten or simplify that process when possible. The long division process can be simplified by a procedure known as *synthetic division* when the divisor is of first degree.

Note that in the process of long division, only the coefficients of the terms are actually used in finding a quotient.

We will show the development of synthetic division by the following example:

Divide $3x^4 - 5x^3 + 9x + 5$ by $(x - 2)$. First by long division we obtain

$$
\begin{array}{r}
3x^3 + x^2 + 2x + 13 \\
x - 2 \overline{\smash{)}3x^4 - 5x^3 + 0x^2 + 9x + 5} \\
\underline{3x^4 - 6x^3} \\
x^3 + 0x^2 + 9x + 5 \\
\underline{x^3 - 2x^2} \\
2x^2 + 9x + 5 \\
\underline{2x^2 - 4x} \\
13x + 5 \\
\underline{13x - 26} \\
31
\end{array}
$$

The quotient is $3x^3 + x^2 + 2x + 13$ and the remainder is 31.

If we are careful to keep like powers of the variable under one another and replace missing terms by 0, we can eliminate some of the unnecessary detail in the division.

First, we note that the repeated terms $3x^4$, x^3, $2x^2$, and $13x$ could be eliminated. Also it is not necessary to keep bringing down terms from the dividend. Thus we could write the division as

$$
\begin{array}{r}
3x^3 + x^2 + 2x + 13 \\
x - 2 \overline{\smash{\big)}\, 3x^4 - 5x^3 + 0x^2 + 9x + 5} \\
\underline{-\ 6x^3 } \\
x^3 \\
\underline{-\ 2x^2 } \\
2x^2 \\
\underline{-\ 4x } \\
13x \\
\underline{-\ 26} \\
31
\end{array}
$$

Next, since we are always dividing by $(x - c)$, we could eliminate the x in $(x - 2)$. We could also eliminate the other powers of x. Doing this, we obtain

$$
\begin{array}{r}
3 1 2 13 \\
-2 \overline{\smash{\big)}\, 3\ -5 0 9 5} \\
\underline{-6 } \\
1 \\
\underline{-2 } \\
2 \\
\underline{-\ 4 } \\
13 \\
\underline{-26} \\
31
\end{array}
$$

We may write this in a more compact form as

$$
\begin{array}{r}
3 1 2 13 \\
-2 \overline{\smash{\big)}\, 3\ -5 0 9 5} \\
\underline{-6\ -2\ -\ 4\ -26} \\
1 2 13 31
\end{array}
$$

Now note that the bottom line is the same as the quotient (top line) except for the first entry on the top line and the last entry (remainder) on the bottom line. By eliminating the top line and entering 3 in the bottom line, we further reduce the problem to the form

$$
\begin{array}{r}
-2 \overline{\smash{\big)}\, 3\ -5 0 9 5} \\
\underline{-6\ -2\ -\ 4\ -26} \\
3 1 2 13 31
\end{array}
$$

6.3 SYNTHETIC DIVISION

If we now notice that each entry in line two is obtained by multiplying the preceding entry of line three by -2, we can construct the form just given without referring to the intervening steps.

For instance, if we wish to divide $(5x^2 - 2x + 3)$ by $(x - 3)$, we arrange the problem as follows:

$$-3\,\underline{/\,5\ -2\ \ \ 3}$$

We first place 5 in the first position in row three. We then multiply it by -3 obtaining

$$\begin{array}{r} -3\,\underline{/\,5\ -\ 2\ \ \ 3} \\ \underline{-15} \\ 5 \end{array}$$

Then we subtract -15 from -2 getting 13 in the second position of row three. We then multiply it by -3 and enter the product in the next position in row two.

$$\begin{array}{r} -3\,\underline{/\,5\ -\ 2\ \ \ \ 3} \\ \underline{-15\ -39} \\ 5\ \ \ \ 13 \end{array}$$

Subtracting again, we obtain a remainder of 42.

$$\begin{array}{r} -3\,\underline{/\,5\ -\ 2\ \ \ \ 3} \\ \underline{-15\ -39} \\ 5\ \ \ \ 13\ \ \ 42 \end{array}$$

Therefore we have

$$5x^2 - 2x + 3 = (x - 3)(5x + 13) + 42$$

We should also recall that the remainder tells us that

$$f(3) = 42$$

One final improvement may be made on this form. It is possible to avoid the subtraction in this form by using c instead of $-c$ when dividing by $(x - c)$. This, in effect, changes all the signs of the entries in the second row. Now, if we wish to divide $(5x^2 - 2x + 3)$ by $(x - 3)$, we write

$$\begin{array}{r} 3\,\underline{/\,5\ -\ 2\ \ \ \ 3} \\ \underline{15\ \ \ 39} \\ 5\ \ \ \ 13\ \ \ 42 \end{array}$$

Notice that we add rather than subtract. This process is called *synthetic division*.

CAUTION: **Be careful of the sign of the number you divide by. For example, if you are dividing by $(x - 3)$, use 3 in your synthetic division. Likewise, if you are dividing by $(x + 5)$, use -5 in your synthetic division.**

EXAMPLE

Using synthetic division, find the quotient and remainder if $2x^4 + 5x^3 - 2x^2 + x - 8$ is divided by $(x + 3)$. Notice that $c = -3$. Thus we divide by -3.

$$
\begin{array}{r|rrrrr}
-3 & 2 & 5 & -2 & 1 & -8 \\
 &   & -6 & 3 & -3 & 6 \\
\hline
 & 2 & -1 & 1 & -2 & -2
\end{array}
$$

So $\quad 2x^4 + 5x^3 - 2x^2 + x - 8 = (x + 3)(2x^3 - x^2 + x - 2) - 2$

EXAMPLE

If $f(x) = 3x^5 - 38x^3 + 5x^2 + 4$, find $f(4)$ by synthetic division. By the remainder theorem, $f(4)$ is the remainder when $f(x)$ is divided by $(x - 4)$. We divide by synthetic division obtaining

$$
\begin{array}{r|rrrrrr}
4 & 3 & 0 & -38 & 5 & 0 & 4 \\
 &   & 12 & 48 & 40 & 180 & 720 \\
\hline
 & 3 & 12 & 10 & 45 & 180 & 724
\end{array}
$$

Therefore $f(4) = 724$.

Exercise 6.3.1

Find the quotient and remainder using synthetic division.

1. $(x^3 - 4x^2 + 2x + 4) \div (x - 3)$
2. $(x^3 - 8) \div (x - 2)$
3. $(2x^4 + x^3 - 1) \div (x + 2)$
4. $(x^6 + x^4 - x) \div (x - 1)$
5. $(x^6 + 4x^5 + 2x^3 + 7x^2 - 4x + 1) \div (x + 4)$

Use synthetic division to solve the following problems:

6. Find $f(2)$ if $f(x) = 2x^3 - x^2 + x + 4$.
7. Find $f(4)$ if $f(x) = x^3 - 4x^2 + x - 13$.
8. Find $g(-1)$ if $g(x) = 3x^3 + 5x^2 + x - 1$.

9. Find $h(-3)$ if $h(x) = 2x^5 + 5x^4 + 10x^2 + 2$.
10. Find $g(5)$ if $g(x) = x^4 - 2x^3 - 17x^2 + 8x + 10$.
11. Is 2 a zero of $f(x) = 3x^4 - 5x^3 + x + 10$?
12. Is 7 a zero of $g(x) = x^3 - 6x^2 - 11x + 28$?
13. Is -1 a zero of $h(x) = 4x^5 - 3x^3 + 1$?
14. Is $\frac{1}{3}$ a zero of $f(x) = 3x^5 + 5x^4 - 2x^3 + 3x^2 - 4x + 1$?

Although we have been applying the properties of polynomials to real numbers, we should not forget that these principles also hold for the field of complex numbers. To illustrate this, use the process of synthetic division to solve the following:

15. Is $(1 + i)$ a zero of $f(x) = 3x^3 - 5x^2 + 4x + 2$?
16. Is $(1 - i)$ a zero of $f(x) = 3x^3 - 5x^2 + 4x + 2$?

6.4 ZEROS OF POLYNOMIALS WITH COMPLEX COEFFICIENTS

Goals

Upon completion of this section you should be able to:

1. Determine the number of zeros a given polynomial has.
2. Find a polynomial having a given set of zeros.

* *

The factor theorem in Sec. 6.2 shows that there is a definite connection between the zeros of a polynomial and the factors of a polynomial. In fact, if we can find a number c such that $f(c) = 0$, then we have a factor $(x - c)$. However, zeros of a polynomial are usually very difficult to find. In special cases, such as linear or quadratic polynomials, we have developed methods of finding zeros. But in other cases, such as $f(x) = x^5 - 3x^4 + 5x^3 - 2x + 10$, there are no obvious zeros and there are no special formulas, such as the quadratic formula, that will yield zeros. Even though it is difficult to locate zeros of polynomials, we can make some headway concerning the theory of such zeros. The following theorem is basic to the development of the theory.

Fundamental Theorem of Algebra   If $f(x)$ is a polynomial of degree greater than zero, with complex coefficients, then $f(x)$ has at least one complex zero.

The proof of this theorem is beyond the scope of this text, but we will accept it without proof and proceed to use it in the development of our theory.

Theorem   If $f(x)$ is a polynomial of degree $n > 0$ with complex coefficients, then $f(x)$ has exactly n zeros. (Note that this does not imply n different zeros. If a zero occurs more than once, it is counted each time.)

Proof   By the fundamental theorem of algebra, $f(x)$ has at least one zero. We designate it as c_1. Now by the factor theorem, $(x - c_1)$ is a factor of $f(x)$ and hence

$$f(x) = (x - c_1) Q_1(x)$$

The degree of $Q_1(x)$ is $n - 1$. Why?

If $n - 1 = 0$, then $n = 1$ and $f(x)$ has one zero, but if $n - 1 > 0$, then by the fundamental theorem, $Q_1(x)$ has a zero, call it c_2, and by the factor theorem

$$Q_1(x) = (x - c_2) Q_2(x)$$

or

$$f(x) = (x - c_1)(x - c_2) Q_2(x)$$

If we repeat the argument until the degree of $Q_n(x) = 0$, then we will have n zeros.

The theorem states "exactly n zeros," so we must now show that $f(x)$ can have no more than n zeros. We can do this by the indirect method of proof. We assume that $f(x)$ of degree $n > 0$ has more than n zeros. This means that $f(x)$ when factored in the form

$$f(x) = (x - c_1)(x - c_2)(x - c_3) \cdots$$

has more than n factors. Each factor $(x - c_i)$ is of degree 1. The product of these factors would yield a polynomial whose degree would be the sum of the degrees of the individual factors. Thus $f(x)$ would be of degree greater than n. This is a contradiction of the hypothesis which states that $f(x)$ is of degree n. Therefore $f(x)$ cannot have more than n zeros.

We have thus shown that $f(x)$ must have exactly n zeros.

We have assumed the fundamental theorem to be true and used it to show that we can determine the number of zeros any polynomial will have. Finding the zeros can be easier when we know exactly how many we must find. We know that $f(x) = 2x^3 - 7x^2 + x - 3$ has three zeros, for instance.

Furthermore, we know that if some polynomial $g(x)$ of degree 3 has zeros 2, -3, 1, then it has factors $(x - 2)$, $(x + 3)$, and $(x - 1)$ and that $g(x) = (x - 2)(x + 3)(x - 1)$ or $g(x) = x^3 - 7x + 6$.

EXAMPLE

Find a third-degree polynomial that has zeros 3, i, $-i$. We can write
$$f(x) = (x - 3)(x - i)(x + i)$$
and by multiplying obtain
$$f(x) = x^3 - 3x^2 + x - 3$$

Exercise 6.4.1

1. How many zeros does each of the following polynomials have?

    a. $f(x) = 4x^2 - 2x + 1$ b. $f(x) = x^2$

    c. $f(x) = x^3 - 1$ d. $f(x) = 3x^5 - (2 + i)x^3 + ix - 2$

    e. $f(x) = 3x + 3 + i$ f. $f(x) = (x - 1)(x - 1)(x - 1)(x - 1)$

2. Find a third-degree polynomial that has zeros -1, 1, 2.
3. Find a third-degree polynomial that has zeros 1, $\sqrt{2}$, $3\sqrt{2}$.
4. Find a third-degree polynomial that has zeros -2, 5, i.
5. Find a third-degree polynomial that has zeros i, $-i$, $2i$.
6. Find a third-degree polynomial that has zeros $1 + i$, $1 - i$, 3.
7. Find a fourth-degree polynomial that has zeros 2, -2, 5, -5.
8. Find a fourth-degree polynomial that has zeros $1 + i$, $1 - i$, i, $-i$.
9. Find the zeros of $f(x) = x^2 - x - 30$.
10. Find the zeros of $f(x) = 2x^2 + x - 6$.
11. Find the zeros of $g(x) = x^4 - 1$.
12. If $f(x) = x^3 - 3x^2 - 6x + 8$ has 1 as one of its zeros, find the other zeros. [*Hint:* Use synthetic division to factor $f(x)$.]

13. If $f(x) = x^3 + 4x^2 - 36x - 144$ has -4 as one of its zeros, find the other zeros.

14. If $f(x) = x^3 - ix^2 - x + i$ has i as one of its zeros, find the other zeros.

15. If $g(x) = x^3 + 3x^2 + x + 3$ has -3 as one of its zeros, find the other zeros.

16. If $f(x) = 2x^3 - x^2 - 4x + 2$ has $\frac{1}{2}$ as one of its zeros, find the other zeros.

17. Use the polynomial $f(x) = x^3 - 1$ to find the three cube roots of 1. [*Hint:* Inspection should give you one zero. Then use synthetic division to factor $f(x)$.]

18. If the same zero appears in a polynomial in n factors, then we say the zero is of multiplicity n. For example, if $f(x) = (x - 5)(x - 5)(x + 1)$, we say that the zero 5 is of multiplicity 2. Find the polynomial of degree 3 that has 1 as a zero of multiplicity 3.

6.5 ZEROS OF POLYNOMIALS WITH REAL COEFFICIENTS

Goals

Upon completion of this section you should be able to:

1. Make use of the fact that complex roots occur in conjugate pairs.
2. Determine the possible number of positive and negative real zeros of a polynomial.

* *

Since the set of real numbers is a subset of the set of complex numbers, any theorem concerning complex numbers will also hold true in the real numbers. In this section we wish to concentrate on zeros of polynomials with real coefficients.

Theorem

If a polynomial $f(x)$ of degree $n > 0$ with real coefficients has as a zero $(a + bi)$, then $(a - bi)$ is also a zero.

Proof We will represent $f(x)$ by
$$f(x) = a_0 x^n + a_1 x^{n-1} + a_2 x^{n-2} + \cdots + a_n$$
and replacing x by $(a + bi)$, we have
$$f(a + bi) = a_0(a + bi)^n + a_1(a + bi)^{n-1} + \cdots + a_n$$
If we expand all of the terms and simplify, we will obtain some real numbers and some imaginary numbers. In fact all terms containing even

powers of i will be real and all terms containing odd powers of i will be imaginary. If we designate the algebraic sum of all the real numbers as P and the sum of all the imaginary numbers as Qi, then we have

$$f(a + bi) = P + Qi$$

but since $(a + bi)$ is given as a zero by the hypothesis, we have

$$f(a + bi) = 0$$

and hence
$$P + Qi = 0$$

Since a complex number $P + Qi$ can equal zero only if the real part and the imaginary part are both zero, we have $P = 0$ and $Q = 0$.

If we replace i with $-i$ in the expansion of $(a + bi)$, those terms with even powers of i will not be affected and those terms with odd powers of i will have a sign opposite their original sign. Therefore

$$f(a - bi) = P - Qi$$

but since $P = 0$ and $Q = 0$

$$f(a - bi) = 0$$

and thus $(a - bi)$ is a zero of $f(x)$.

Theorem **Every polynomial $f(x)$ with real coefficients and of odd degree has at least one real zero.**

Proof We are given that $f(x)$ is of degree n and n is an odd number. We know from the previous section that $f(x)$ has n zeros. Since complex zeros that are not real must occur in pairs, $(a + bi)$ and $(a - bi)$, it follows that $f(x)$ must have at least one real zero.

Note that if $f(x)$ has degree n where n is even, then all zeros could be nonreal.

A polynomial $f(x)$ with real coefficients is said to have a *sign variation* if, when arranged in descending powers of the variable x, the sign of a term is different from the sign of the preceding term. (We ignore missing terms in counting the sign variations.) For instance,

$$f(x) = x^5 + 3x^4 - 2x^3 - x + 5$$

has two sign variations—one between the second and third terms, and one between the fourth and fifth terms.

We now state, without proof, a useful theorem called *Descartes' Rule of Signs*.

Theorem   If $f(x)$ is a polynomial with real coefficients, the number of positive real zeros of $f(x)$ is either equal to the number of sign variations of $f(x)$ or less than the number of sign variations by a positive even integer.

Also, the number of negative real zeros of $f(x)$ is either equal to the number of sign variations of $f(-x)$ or less than the number of sign variations of $f(-x)$ by a positive even integer.

EXAMPLE   Determine the possible number of positive and negative real zeros of $f(x) = 3x^5 - x^4 + 3x^3 - 2x^2 - x - 1$.

$f(x)$ has three sign variations. Therefore the number of positive real zeros is either three or one.

$$f(-x) = -3x^5 - x^4 - 3x^3 - 2x^2 + x - 1$$

has two sign variations, and therefore the number of negative real zeros is 2 or 0.

EXAMPLE   Show by Descartes' Rule of Signs that $f(x) = x^3 + 2x + 3$ has only one real zero.

Descartes' Rule of Signs can be used to check for positive and negative real zeros. However, since 0 is neither positive nor negative, we must check to see if 0 is a zero of $f(x)$. Checking this, we find

$$f(0) = 3$$

which indicates that 0 is not a zero.

Now continuing with Descartes' Rule of Signs, we see that $f(x)$ has no sign variation and therefore no positive real zeros.

$$f(-x) = -x^3 - 2x + 3$$

has one sign variation and therefore $f(x)$ has one negative real zero. Since we know that $f(x)$ is of degree 3 and therefore has three zeros, the other two cannot be real. Hence, $f(x)$ has one negative real zero and two nonreal zeros.

Exercise 6.5.1

1. Given $f(x) = 3x^4 + 2x^3 + x^2 - 2x - 4$

    a. What are the possible numbers of positive real zeros of $f(x)$?

    b. What are the possible numbers of negative real zeros of $f(x)$?

c. Is 0 a zero of $f(x)$? Explain.

2. Given $f(x) = 2x^4 - 5x^3 + x^2$

   a. How many positive real zeros could $f(x)$ have?

   b. How many negative real zeros could $f(x)$ have?

   c. Is 0 a zero of $f(x)$? Explain.

3. What condition is necessary and sufficient for any polynomial $f(x)$ to have 0 as a zero?

4. Show that $f(x) = x^5 + 2x^3 + x - 3$ has only one real zero.

5. Show that $g(x) = x^4 + x^2 + 2x - 3$ has one positive real zero and one negative real zero.

6. Show that $h(x) = x^4 - 5x^3 + 2x^2 - 3x + 1$ has no negative zeros.

7. Show that $f(x) = x^6 + 3x^4 + 5x^2 + 1$ has no real zeros.

8. How many real zeros does $f(x) = x^6 - 5$ have?

9. How many zeros of $f(x) = x^4 + x^2 + x$ are not real?

10. Given $f(x) = x^n + 1$

    a. If n is odd, how many real zeros does $f(x)$ have?

    b. If n is even, how many real zeros does $f(x)$ have?

6.6 ZEROS OF POLYNOMIALS WITH RATIONAL OR INTEGRAL COEFFICIENTS

Goals

Upon completion of this section you should be able to:

1. Give all possible rational zeros of a polynomial.
2. Find all zeros of a given polynomial.

* *

A rational number has been defined as the ratio of integers (that is, $\frac{a}{b}$, where a and b are integers and $b \neq 0$). If $f(x)$ is a polynomial having some nonintegral rational coefficients (that is, fractions), then for $f(x) = 0$ the polynomial can easily be changed to one having integral coefficients by multiplying by a common denominator. This would yield a different polynomial but would not affect the zeros of the original polynomial. For instance,

$$f(x) = \frac{2}{3}x^3 + x^2 - \frac{1}{2}x + \frac{1}{3}$$

would have the same zeros as

$$g(x) = 4x^3 + 6x^2 - 3x + 2$$

obtained by multiplying all terms of $f(x)$ by 6.

We will, therefore, in this section, deal only with the theory of polynomials that have integral coefficients.

Theorem   If the polynomial $f(x) = a_0 x^n + a_1 x^{n-1} + a_2 x^{n-2} + \cdots + a_n$, where $a_0 \neq 0$, has a rational zero $\frac{p}{q}$ (where $\frac{p}{q}$ is in reduced form), then p is an exact divisor of a_n and q is an exact divisor of a_0.

Proof   Since $f\left(\frac{p}{q}\right) = 0$, we have

$$a_0 \left(\frac{p}{q}\right)^n + a_1 \left(\frac{p}{q}\right)^{n-1} + \cdots + a_{n-1}\left(\frac{p}{q}\right) + a_n = 0$$

Multiplying by q^n gives

$$a_0 p^n + a_1 p^{n-1} q + a_2 p^{n-2} q^2 + \cdots + a_{n-1} p q^{n-1} + a_n q^n = 0$$

or

$$a_0 p^n + a_1 p^{n-1} q + a_2 p^{n-2} q^2 + \cdots + a_{n-1} p q^{n-1} = -a_n q^n$$

Factoring p on the left gives

$$p\left(a_0 p^{n-1} + a_1 p^{n-2} q + \cdots + a_{n-1} q^{n-1}\right) = -a_n q^n$$

Since only integers are involved on each side of the equation, we conclude that p is a factor of the integer $-a_n q^n$. We know that p has no factor in common with q (since $\frac{p}{q}$ was in reduced form), and therefore p must be an exact divisor of a_n.

A similar argument can be used to show that q must be an exact divisor of a_0.

EXAMPLE

List all possible rational zeros of $f(x) = 5x^3 + 3x^2 + 2x - 6$.

The divisors of 5 are ± 1 and ± 5, and the divisors of 6 are $\pm 1, \pm 2, \pm 3,$ and ± 6. By the theorem just proved, if $\dfrac{p}{q}$ is a zero, then p must be an exact divisor of 6 and q must be an exact divisor of 5. Therefore, the possible rational zeros are

$$\left\{ \pm \frac{1}{5}, \pm \frac{2}{5}, \pm \frac{3}{5}, \pm 1, \pm \frac{6}{5}, \pm 2, \pm 3, \pm 6 \right\}$$

CAUTION: Note that the theorem does not guarantee that such zeros exist, but simply limits the possible rational zeros should they exist.

Two obvious corollaries of the theorem are:

Corollary 1 An integral zero of the polynomial $f(x)$ must be an exact divisor of the constant term a_n.

Corollary 2 If the polynomial $f(x)$ has the leading coefficient $a_0 = 1$, then any rational zero is an integer.

The student will be asked to prove these corollaries as exercises.

EXAMPLE

Find all zeros of $f(x) = 2x^3 - 5x^2 + x + 2$.

From the work in this chapter we know

a. $f(x)$ has at least one real zero. (Why?)
b. the number of positive real zeros is 2 or 0. (Why?)
c. the number of negative real zeros is 1. (Why?)
d. the possible rational zeros are in the set

$$\left\{ \pm \frac{1}{2}, \pm 1, \pm 2 \right\}$$

Having this information, it is logical to check to see if -1, -2, or $-\dfrac{1}{2}$ is a zero. (Why?) We may do this by substituting each of these values for x in $f(x)$ or by using synthetic division. The advantage of using synthetic division is that when we find a zero, we will also have the coefficients of the other factor of the polynomial.

First we check −1.

$$\begin{array}{r|rrrr} -1 & 2 & -5 & 1 & 8 \\ & & -2 & 7 & -8 \\ \hline & 2 & -7 & 8 & -6 \end{array}$$

Therefore −1 is not a zero. Checking −2 we obtain

$$\begin{array}{r|rrrr} -2 & 2 & -5 & 1 & 2 \\ & & -4 & 18 & -38 \\ \hline & 2 & -9 & 19 & -36 \end{array}$$

Thus −2 is not a zero. Next we try $-\frac{1}{2}$.

$$\begin{array}{r|rrrr} -\frac{1}{2} & 2 & -5 & 1 & 2 \\ & & -1 & 3 & -2 \\ \hline & 2 & -6 & 4 & 0 \end{array}$$

We see here that $-\frac{1}{2}$ is a zero and we may write

$$f(x) = \left(x + \frac{1}{2}\right)(2x^2 - 6x + 4)$$

Since $2x^2 - 6x + 4 = 2(x-1)(x-2)$, we now have

$$f(x) = 2\left(x + \frac{1}{2}\right)(x-1)(x-2)$$

Thus the zeros of $f(x)$ are $\left\{-\frac{1}{2}, 1, 2\right\}$.

CAUTION: Note that if no zero of $f(x)$ had been rational, this method would not have given us the answers. Of course, if enough of the zeros are rational to allow us to get to a quadratic factor, we may solve the quadratic to give us the remaining two zeros whether they are rational or not.

In the next section we will deal with a method of approximating irrational zeros.

6.6 ZEROS OF POLYNOMIALS: RATIONAL/INTEGRAL COEFFICIENTS

Exercise 6.6.1

Using the theory developed in this chapter, find all zeros of the following polynomials:

1. $f(x) = x^3 - 2x^2 - x + 2$
2. $f(x) = x^3 - 4x^2 + x + 6$
3. $f(x) = 2x^3 - 3x^2 - 11x + 6$
4. $f(x) = 2x^3 + 3x^2 - 14x - 21$
5. $f(x) = 18x^3 - 9x^2 - 5x + 2$
6. $f(x) = 16x^3 + 4x^2 - 4x - 1$
7. $f(x) = 2x^3 + 7x^2 + 2x - 6$
8. $f(x) = x^3 + 5x^2 + x + 5$
9. $f(x) = 2x^4 + 5x^3 - 5x^2 - 5x + 3$
10. $f(x) = x^4 - 2x^3 - 13x^2 + 38x - 24$
11. $f(x) = 3x^4 - 11x^3 + 9x^2 + 13x - 10$
12. $f(x) = x^4 + x^3 - 5x^2 - 15x - 18$
13. $f(x) = x^5 - 23x^4 + 160x^3 - 281x^2 - 257x - 440$
14. Prove Corollary 1 from this section.
15. Prove Corollary 2 from this section.

6.7 SOME THEOREMS ON THE BOUNDARIES FOR THE ZEROS OF POLYNOMIALS

Goals

Upon completion of this section you should be able to:

1. Determine upper and lower bounds for real roots of a polynomial.
2. Locate the approximate position of irrational roots of a polynomial.

* *

It was previously stated that zeros of polynomials are difficult to find except in special cases. No general method exists for finding the zeros of a fifth- or higher-degree polynomial. The theorems on zeros from the previous sections help us to find zeros in many situations. In this section we wish to investigate some other theorems which will aid in our search for zeros and also discuss some methods of estimating irrational zeros.

Theorem If $f(x) = a_0 x^n + a_1 x^{n-1} + \cdots + a_n$ and if a_k is the coefficient of greatest absolute value, then no positive zero can exceed $\left|\dfrac{a_k}{a_0}\right| + 1$.

This theorem is stated without proof.

EXAMPLE Determine the upper limit for the positive zeros of $f(x)$.
$$f(x) = 8x^7 + 3x^6 - 15x^3 + x - 12$$
Using the theorem, we know that we cannot have a positive zero greater than
$$\left|\frac{-15}{8}\right| + 1 = 2\frac{7}{8}$$
Thus if we were searching for zeros for $f(x)$, we would not bother to check the possibilities 3, 4, 6, or 12.

To find *bounds* for negative zeros (i.e., points between which negative zeros occur), it is only necessary to find $f(-x)$ and use the bounds for positive zeros with the opposite sign. For the theorem just given, since coefficients of $f(-x)$ would have the same absolute value, the lower bounds of negative zeros are represented by $-\left(\left|\dfrac{a_k}{a_0}\right| + 1\right)$.

Theorem If, in the synthetic division of a polynomial $f(x)$ with real coefficients by $(x - a)$ where $a > 0$, all the numbers in the third row are positive, then $f(x)$ has no real zero larger than a.

Theorem If, in the synthetic division of a polynomial $f(x)$ with real coefficients by $(x - a)$ where $a < 0$, the numbers in the third row are alternately positive and negative, then $f(x)$ has no real zero less than a.

Theorem If for the polynomial $f(x)$ with real coefficients, $f(a)$ and $f(b)$ are opposite in sign, then $f(x)$ has a real zero between a and b.

A graphical representation, assuming that a polynomial forms a continuous curve, illustrates this last theorem.

6.7 BOUNDARIES FOR THE ZEROS OF POLYNOMIALS

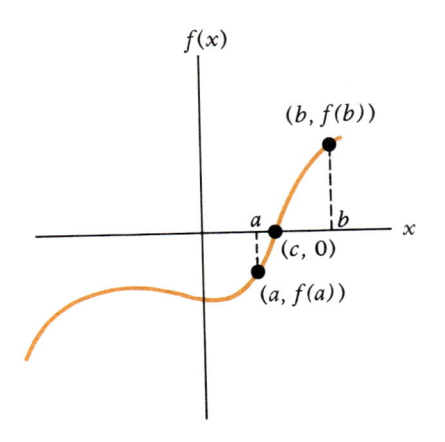

A restatement of the theorem in terms of the graph would be: "If $f(a)$ is below the x-axis and $f(b)$ is above the x-axis, then the curve must cross the x-axis between a and b." Of course, a zero of $f(x)$ is where $f(x) = 0$ or where the curve crosses the x-axis. It is noted that if c is a zero of $f(x)$, then the point on the polynomial curve would be $(c,0)$.

EXAMPLE

Discuss the zeros of $f(x) = x^3 + 4x^2 + x - 3$.

From the theorems in the chapter we know

a. $f(x)$ has three zeros. (Why?)
b. the possible rational zeros are $\{\pm 1, \pm 3\}$. (Why?)
c. $f(x)$ has one positive real zero. (Why?)
d. $f(x)$ has two or no negative real zeros. (Why?)
e. an upper bound of the zeros is 5 and a lower bound is -5. (Why?)

Since we know there is one positive real zero, we will first try to locate it. We first note that

$$f(0) = -3$$

We now find $f(1)$ by using synthetic division.

$$\underline{1\,/\ \ 1 \quad 4 \quad 1 \quad -3}$$
$$\ \ 1 \quad 5 \quad \ \ 6$$
$$1 \quad 5 \quad 6 \quad \ \ 3$$

Thus $f(1) = 3$.

Now since $f(0) < 0$ and $f(1) > 0$, there must be a real zero between 0 and 1. Also, since any rational zero must be an integer [see (b) above], we know that this zero must be irrational. (We should note here that all the numbers in the third row are positive and, therefore, there are no real zeros greater than 1. However, we know that there is only one positive real zero, and we have already located it.)

Next we check for negative real zeros. We know that if there are any, there will be two of them. First we check -1 by synthetic division.

$$\begin{array}{r|rrrr} -1 & 1 & 4 & 1 & -3 \\ & & -1 & -3 & 2 \\ \hline & 1 & 3 & -2 & -1 \end{array}$$

Thus $f(-1) = -1$. Similarly we check -2.

$$\begin{array}{r|rrrr} -2 & 1 & 4 & 1 & -3 \\ & & -2 & -4 & 6 \\ \hline & 1 & 2 & -3 & 3 \end{array}$$

Thus $f(-2) = 3$. We note that $f(-1) < 0$ and $f(-2) > 0$. Hence there is a real zero between -2 and -1 and it must also be irrational.

The next check -3.

$$\begin{array}{r|rrrr} -3 & 1 & 4 & 1 & -3 \\ & & -3 & -3 & 6 \\ \hline & 1 & 1 & -2 & 3 \end{array}$$

Thus $f(-3) = 3$. Next, checking -4, we obtain

$$\begin{array}{r|rrrr} -4 & 1 & 4 & 1 & -3 \\ & & -4 & 0 & -4 \\ \hline & 1 & 0 & 1 & -7 \end{array}$$

Hence $f(-4) = -7$. We see that $f(-3) > 0$ and $f(-4) < 0$, thus there is a real zero between -4 and -3 and it is also irrational.

We thus conclude that there are three irrational zeros. One of them is in the interval $(-4, -3)$, another is in the interval $(-2, -1)$, and the third is in the interval $(0, 1)$.

Attempting to estimate irrational zeros seems to be the next logical step. Let us use the example just discussed and approximate the zero in the interval $(-2, -1)$.

6.7 BOUNDARIES FOR THE ZEROS OF POLYNOMIALS

We found in the example that
$$f(-1) = -1 \quad \text{and} \quad f(-2) = 3$$
The curve representing $f(x)$ must intersect the x-axis somewhere between -1 and -2. If we approximate the curve between the points $(-2,3)$ and $(-1,-1)$ by a line segment, we would obtain the following figure.

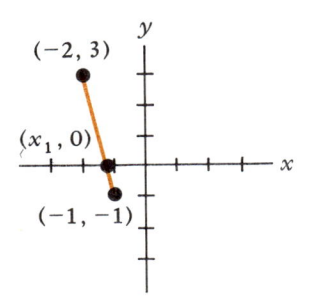

A line segment containing the points $(-2,3)$ and $(-1,-1)$ will intersect the x-axis close (in some sense) to the point where $f(x)$ intersects the x-axis, $(x_1, 0)$. Using the two-point form of a straight line

$$\frac{y - y_1}{x - x_1} = \frac{y_2 - y_1}{x_2 - x_1}$$

gives
$$\frac{y + 1}{x + 1} = \frac{3 + 1}{-2 + 1}$$

Now if we substitute $(x_1, 0)$ for (x,y), we obtain

$$\frac{1}{x_1 + 1} = \frac{4}{-1}$$

or
$$x_1 = -\frac{5}{4} = -1.25$$

This is an approximation of the zero between -2 and -1. If we wished a better approximation, we would continue by finding $f(-1.25)$ using synthetic division.

$$\begin{array}{r|rrrr}
-1.25 & 1 & 4 & 1 & -3 \\
 & & -1.25 & -3.4375 & 3.0469 \\
\hline
 & 1 & 2.75 & -2.4375 & 0.0469
\end{array}$$

216   POLYNOMIALS

Thus $f(-1.25) = 0.0469$. Since $f(-1) < 0$ and $f(-1.25) > 0$, the irrational zero must be between -1 and -1.25.

If we use the two points $(-1,-1)$ and $(-1.25, 0.0469)$ and join them with a line segment, that line segment will intersect the x-axis close (in some sense) to the point where $f(x)$ intersects the x-axis, $(x_2, 0)$. Again we have

$$\frac{y+1}{x+1} = \frac{0.0469 + 1}{-1.25 + 1}$$

or

$$\frac{y+1}{x+1} = \frac{1.0469}{-0.25}$$

If we substitute $(x_2, 0)$ for (x,y) and simplify, we obtain

$$x_2 = -1.2388$$

which is our second approximation of the zero.

This method can be continued to obtain the approximation of the zero to any degree of accuracy desired. A calculator would aid greatly in the computation.

Exercise 6.7.1

Locate the intervals for the real zeros of the following polynomials:

1. $f(x) = x^3 + x^2 - 2x - 1$
2. $f(x) = x^3 - 7x + 5$
3. $f(x) = x^3 - 33x + 20$
4. $f(x) = x^3 - x^2 - 2$
5. $f(x) = x^3 - x^2 + x - 4$

Find, to two decimal places, the indicated real zero in each of the following polynomials.

6. $f(x) = x^3 + x^2 - 10x + 4$, the zero between 0 and 1.
7. $f(x) = x^3 + 3x^2 - 9x - 3$, the zero between 2 and 3.
8. $f(x) = x^4 - 5x^3 + 2x^2 + x + 7$, the zero between 1 and 2.
9. $f(x) = 4x^3 + 6x^2 + 3x + 5$, the zero between -2 and -1.
10. $f(x) = 2x^3 - 7x^2 - 16x + 33$, the zero between -3 and -2.

Find all zeros of the following polynomials to two decimal places:

11. $f(x) = 2x^3 - 4x^2 - 10x - 3$
12. $f(x) = 8x^3 + 12x^2 - 50x + 29$

6.8 PARTIAL FRACTIONS

Goals

Upon completion of this section you should be able to:

1. Express an algebraic fraction as the sum of partial fractions.

* *

In Section 1.5 we reviewed the process of adding or subtracting two or more fractions to obtain a single reduced fraction. In integral calculus it is sometimes desirable to reverse this process. That is, find the individual fractions which sum to a given fraction.

Suppose we have a given fraction with a numerator and denominator that are both polynomials, and we wish to separate it into a sum of partial fractions (i.e., separate $\frac{f(x)}{g(x)}$ into partial fractions). Our success in doing so depends on two things.

1. The degree of $f(x)$ must be less than the degree of $g(x)$. If this is not the case, we can divide $f(x)$ by $g(x)$ using long division and then work with the remaining fractions. (See Sec. 6.2.)
2. The factors of the denominator $g(x)$ must be known.

Problems with partial fractions can be classified into three basic types, and we will, by example, show how to solve each type.

Type I. If $g(x)$, the denominator, has prime linear factors $(x - r_1)$, $(x - r_2), (x - r_3), \cdots, (x - r_n)$, all different, then the fraction

$$\frac{f(x)}{g(x)} = \frac{A}{x - r_1} + \frac{B}{x - r_2} + \frac{C}{x - r_3} + \cdots + \frac{D}{x - r_n}$$

Type II. If $g(x)$, the denominator, has linear factors $(x - r)^n$, then

$$\frac{f(x)}{g(x)} = \frac{A}{x - r} + \frac{B}{(x - r)^2} + \frac{C}{(x - r)^3} + \cdots + \frac{D}{(x - r)^n}$$

Type III. If $g(x)$, the denominator, has quadratic factors $(x^2 + bx + c)^n$, then

$$\frac{f(x)}{g(x)} = \frac{Ax + B}{x^2 + bx + c} + \frac{Cx + D}{(x^2 + bx + c)^2} + \cdots + \frac{Ex + F}{(x^2 + bx + c)^n}$$

Of course, a given fraction might be a combination of any of the three types.

EXAMPLE

Type I: Find partial fractions that sum to $\dfrac{3x - 1}{x^2 + 5x + 6}$.

$$\frac{3x - 1}{x^2 + 5x + 6} = \frac{3x - 1}{(x + 2)(x + 3)} = \frac{A}{x + 2} + \frac{B}{x + 3}$$

Adding the two fractions gives

$$\frac{3x - 1}{(x + 2)(x + 3)} = \frac{A(x + 3) + B(x + 2)}{(x + 2)(x + 3)}$$

By removing parentheses and collecting terms, we get

$$\frac{3x - 1}{(x + 2)(x + 3)} = \frac{(A + B)x + 3A + 2B}{(x + 2)(x + 3)}$$

We now have two fractions with the same denominator, and the numerators of these fractions are polynomials. The fractions will be equal if and only if, in the numerator, coefficients of corresponding powers of x are equal. Therefore

$$A + B = 3 \quad \text{and} \quad 3A + 2B = -1$$

Solving these two equations simultaneously gives $A = -7$ and $B = 10$. Hence

$$\frac{3x - 1}{(x + 2)(x + 3)} = \frac{-7}{(x + 2)} + \frac{10}{(x + 3)}$$

We can check the solution by adding the fractions $-\dfrac{7}{x + 2}$ and $\dfrac{10}{x + 3}$ to get $\dfrac{3x - 1}{(x + 2)(x + 3)}$.

EXAMPLE

Type II: Find partial fractions that sum to $\dfrac{2x + 7}{x^2 + 10x + 25}$.

6.8 PARTIAL FRACTIONS

$$\frac{2x+7}{x^2+10x+25} = \frac{2x+7}{(x+5)^2} = \frac{A}{x+5} + \frac{B}{(x+5)^2}$$
$$= \frac{A(x-5)+B}{(x+5)^2}$$
$$= \frac{Ax+5A+B}{(x+5)^2}$$

Since
$$\frac{2x+7}{(x+5)^2} = \frac{Ax+5A+B}{(x+5)^2}$$

then
$$A = 2 \quad \text{and} \quad 5A + B = 7$$

So
$$A = 2 \quad \text{and} \quad B = -3$$

Hence
$$\frac{2x+7}{(x+5)^2} = \frac{2}{x+5} - \frac{3}{(x+5)^2}$$

Check by adding $\dfrac{2}{x+5}$ and $\dfrac{-3}{(x+5)^2}$ to get $\dfrac{2x+7}{(x+5)^2}$.

EXAMPLE

Type III: Find partial fractions that sum to $\dfrac{3x^2+5x+1}{(x^2+2x+3)^2}$.

$$\frac{3x^2+5x+1}{(x^2+2x+3)^2} = \frac{Ax+B}{x^2+2x+3} + \frac{Cx+D}{(x^2+2x+3)^2}$$
$$= \frac{(Ax+B)(x^2+2x+3)+Cx+D}{(x^2+2x+3)^2}$$
$$= \frac{Ax^3+2Ax^2+3Ax+Bx^2+2Bx+3B+Cx+D}{(x^2+2x+3)^2}$$
$$= \frac{Ax^3+(2A+B)x^2+(3A+2B+C)x+3B+D}{(x^2+2x+3)^2}$$

Equating coefficients of like terms in the numerator gives the four equations

$$A = 0$$
$$2A + B = 3$$
$$3A + 2B + C = 5$$

$$3B + D = 1$$

Solving these simultaneously gives

$$A = 0 \quad B = 3 \quad C = -1 \quad D = -8$$

Therefore

$$\frac{3x^2 + 5x + 1}{(x^2 + 2x + 3)^2} = \frac{3}{x^2 + 2x + 3} - \frac{x + 8}{(x^2 + 2x + 3)^2}$$

This result can be checked by combining the partial fractions.

EXAMPLE Find partial fractions that sum to $\dfrac{x^2 + x - 3}{(x + 1)(x + 2)^2}$.

This example is a combination of Types I and II. Hence

$$\frac{x^2 + x - 3}{(x + 1)(x + 2)^2} = \frac{A}{x + 1} + \frac{B}{x + 2} + \frac{C}{(x + 2)^2}$$

$$= \frac{A(x + 2)^2 + B(x + 1)(x + 2) + C(x + 1)}{(x + 1)(x + 2)^2}$$

$$= \frac{Ax^2 + 4Ax + 4A + Bx^2 + 3Bx + 2B + Cx + C}{(x + 1)(x + 2)^2}$$

$$= \frac{(A + B)x^2 + (4A + 3B + C)x + 4A + 2B + C}{(x + 1)(x + 2)^2}$$

Therefore

$$A + B = 1$$
$$4A + 3B + C = 1$$
$$4A + 2B + C = -3$$

which gives $\quad A = -3 \quad B = 4 \quad C = 1$

Therefore

$$\frac{x^2 + x - 3}{(x + 1)(x + 2)^2} = \frac{-3}{x + 1} + \frac{4}{x + 2} + \frac{1}{(x + 2)^2}$$

EXAMPLE Find partial fractions that sum to $\dfrac{6x - 9}{x^3 - 1}$.

$$\frac{6x - 9}{x^3 - 1} = \frac{6x - 9}{(x - 1)(x^2 + x + 1)}$$

6.8 PARTIAL FRACTIONS

which is a combination of Types I and III. Hence

$$\frac{6x - 9}{(x - 1)(x^2 + x + 1)} = \frac{A}{x - 1} + \frac{Bx + C}{x^2 + x + 1}$$

$$= \frac{Ax^2 + Ax + A + Bx^2 + Cx - Bx - C}{(x - 1)(x^2 + x + 1)}$$

This gives the three equations

$$A + B = 0$$
$$A - B + C = 6$$
$$A - C = -9$$

Solving these gives

$$A = -1 \quad B = 1 \quad C = 8$$

Therefore

$$\frac{6x - 9}{x^3 - 1} = \frac{x + 8}{x^2 + x + 1} - \frac{1}{x - 1}$$

Exercise 6.8.1

Find partial fractions that sum to each of the following:

1. $\dfrac{1}{(x + 1)(x - 2)}$

2. $\dfrac{x}{x^2 + 2x - 3}$

3. $\dfrac{x^2 + 1}{(x - 2)(x - 1)(2x + 1)}$

4. $\dfrac{x}{(x - 1)(x^2 - 5x + 6)}$

5. $\dfrac{x^3 - 1}{x(x + 1)^3}$

6. $\dfrac{1}{(x + 1)^3 (x - 1)}$

7. $\dfrac{x^2 + 2x + 3}{x^3 - x}$

8. $\dfrac{x + 5}{(x - 1)^2 (x + 2)}$

9. $\dfrac{3x^2 + x - 2}{(x - 1)(x^2 + 1)}$

10. $\dfrac{x + 2}{(x^2 + x + 1)(x - 1)}$

11. $\dfrac{3x^2 - x + 1}{(x + 1)(x^2 - x + 3)}$

12. $\dfrac{6x^2 - 3x + 1}{(4x + 1)(x^2 + 1)}$

13. $\dfrac{3x^2 + x + 6}{x^4 + 3x^2 + 2}$

14. $\dfrac{x^5}{(x - 1)(x^2 + 2)^2}$

15. $\dfrac{6x^2 - 15x + 22}{(x + 3)(x^2 + 2)^2}$

16. $\dfrac{x^4 + 2x^3 + 5x^2 + 5x + 3}{(x + 1)(x^2 + 2x + 2)^2}$

CHAPTER REVIEW

1. Using direct substitution in $f(x) = x^5 + 3x^3 - 4x + 5$, find $f(-1)$.
2. Is -3 a zero of $P(x) = x^3 - 2x^2 - 11x + 12$?
3. State the degree of $g(x) = 4x^3 - 2x^2 + 1$.
4. If $f(x) = x^5 - 2x^4 + x - 1$ and $g(x) = 3x^2 + x - 4$, state the degree of $f(x) \cdot g(x)$.
5. If $f(x) = 4x^3 + x + 2$ and $g(x) = (a - 1)x^3 + (b + 1)x^2 + (c - 10)x + 2$, find values for a, b, and c so that $f(x) = g(x)$.

Use long division to find the remainder.

6. $(5x^4 - 2x^3 + x^2 + 2x - 1) \div (x - 2)$
7. $(3x^5 - 12x^3 + x + 4) \div (x + 3)$

Use the remainder theorem to find the remainder.

8. $(x^4 - 5x^3 + 11x^2 + x - 5) \div (x - 1)$
9. $(x^3 - 4x^2 + 3x + 2) \div (x + 3)$

Use the factor theorem to answer Questions 10 and 11.

10. Is $(x - 3)$ a factor of $4x^4 - 11x^3 - 10x - 3$?
11. Is $(x + 2)$ a factor of $x^5 + 2x^4 - x^3 + 4x$?
12. Use synthetic division to divide $x^4 + 3x^3 - 44x^2 + 19x + 6$ by $(x - 5)$.
13. Use synthetic division to divide $x^5 - 30x^3 + 40x^2 - 103$ by $(x + 6)$.
14. If $f(x) = x^6 - 3x^4 + x - 1$, find $f(2)$ using synthetic division.
15. Use synthetic division to find $f(-4)$ if $f(x) = x^4 + 3x^3 + 12x - 1$.
16. Use synthetic division to determine if 6 is a zero of $f(x) = x^4 - 5x^3 - 8x^2 + 11x - 6$.
17. Use synthetic division to determine if -3 is a zero of $g(x) = 5x^4 + 11x^3 + 29x + 6$.
18. Use synthetic division to determine if $(2 + i)$ is a zero of $h(x) = x^3 - 2x^2 - 3x + 10$.
19. How many zeros does $f(x) = 5x^6 - 3x^3 + x^2 - 1$ have?
20. Find a third-degree polynomial that has zeros $-2, 2,$ and 5.
21. Find a third-degree polynomial that has zeros $3, \sqrt{5},$ and $-2\sqrt{5}$.
22. Find a third-degree polynomial that has zeros $1, i,$ and $-i$.

23. Use the polynomial $f(x) = x^3 + 1$ to find the three cube roots of -1.

Using Descartes' Rule of Signs, answer Questions 24 to 29.

24. How many positive real zeros could $f(x) = 6x^4 + x^3 - 2x^2 - x + 1$ have?
25. How many negative real zeros could $f(x) = 6x^4 + x^3 - 2x^2 - x + 1$ have?
26. How many positive real zeros could $f(x) = x^5 + 4x^3 + x - 2$ have?
27. How many negative real zeros could $f(x) = x^5 + 4x^3 + x - 2$ have?
28. How many real zeros does $f(x) = 3x^6 + 2x^4 + 5x^2 + 4$ have?
29. How many real zeros does $f(x) = 2x^5 + 4x^3 + 3x$ have?
30. List all possible rational zeros of $f(x) = 3x^5 - 4x^2 + x - 10$.

Find *all* zeros of the polynomials in Problems 31 to 35.

31. $f(x) = x^3 - 7x^2 + 7x + 15$
32. $f(x) = 4x^3 - 13x + 6$
33. $f(x) = x^4 - 8x^3 + 22x^2 - 24x + 9$
34. $f(x) = 8x^4 + 6x^3 - 15x^2 - 12x - 2$
35. $f(x) = 12x^5 - 44x^4 - 21x^3 + 21x^2 - 6x + 8$

Locate the intervals for the real zeros of the following polynomials:

36. $f(x) = x^3 - 3x^2 - 5$
37. $f(x) = x^4 - 16x^3 + 69x^2 - 70x - 42$

Find, to two decimal places, the indicated real zero in each of the following polynomials:

38. $f(x) = x^3 + 2x^2 - 7x + 1$, the zero between 0 and 1.
39. $f(x) = x^3 - 6x^2 - 8x + 40$, the zero between -3 and -2.
40. Find all real zeros of $f(x) = 8x^3 - 12x^2 + 1$ to two decimal places.

In Problems 41 to 45 find partial fractions that sum to the given fraction.

41. $\dfrac{3x + 29}{x^2 + 3x - 10}$

42. $\dfrac{5x^3 + 20x^2 + 18x + 26}{(x^2 + 2)(x + 2)^2}$

43. $\dfrac{8x^2 + 9x + 22}{x^3 - 8}$

44. $\dfrac{3x^2 + 10x + 11}{(x + 1)^3}$

45. $\dfrac{2x^3 + x^2 - 3x - 10}{(x^2 + 2x + 3)^2}$

PRACTICE TEST

1. If $f(x) = 3x^4 + x^2 - 5x + 6$ and $g(x) = 2x^3 + x^2 - 7x + 1$, state the degree of $f(x) \cdot g(x)$.

2. Find the remainder when $5x^6 - 14x^4 + 11x^3 + x + 4$ is divided by $(x + 2)$.

3. Use the factor theorem to show that $(x + 5)$ is a factor of $x^5 + 6x^4 + 5x^3 + 4x^2 + 18x - 10$.

4. Use synthetic division to determine if -1 is a zero of $f(x) = 3x^5 + 5x^3 + 9x^2 - 1$.

5. Find a third-degree polynomial that has zeros $-3, -1$, and 2.

6. Using Descartes' Rule of Signs, show that $f(x) = 3x^5 - 2x^4 + x^3 - 6x^2 - 1$ has no negative real zeros.

7. List all possible rational zeros of $f(x) = 3x^5 + 2x^4 - 5x^3 + x - 5$.

8. Find *all* zeros of $f(x) = 3x^3 - 5x^2 - 11x - 3$.

9. Find *all* zeros of $f(x) = 2x^3 - x^2 + 8x - 4$.

10. Find *all* zeros of $f(x) = 2x^4 + x^3 - 7x^2 - 3x + 3$.

11. Give the intervals for the real zeros of $f(x) = x^3 + 2x^2 - 15x - 25$.

12. Find partial fractions that sum to give $\dfrac{2x^2 - 34x + 100}{(x + 5)(x - 3)^2}$.

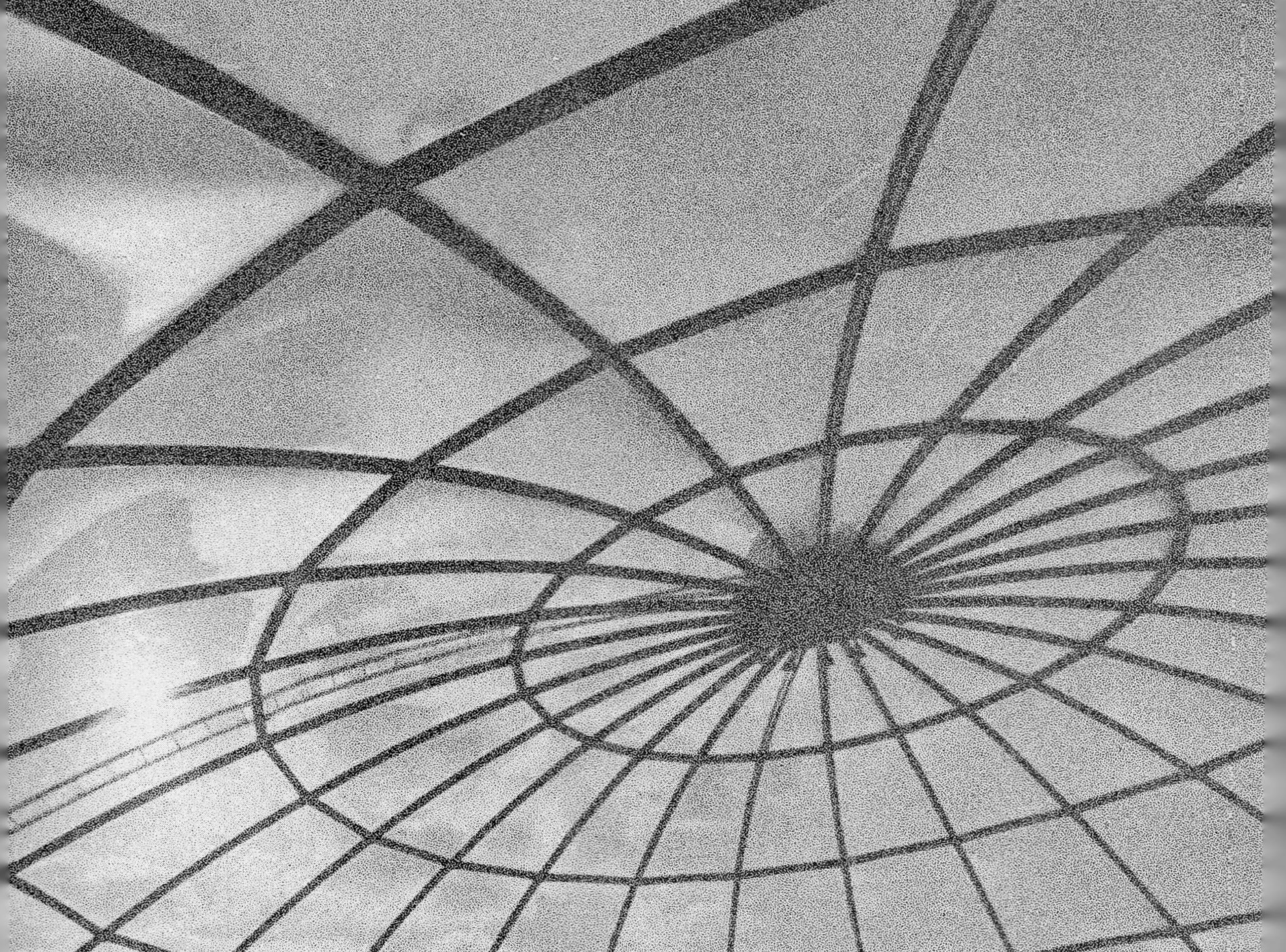

Sequences and Series
7

7

Most of our work in algebra involves the set of real numbers. We have also done some work involving the set of complex numbers. These sets of numbers were developed rather late on the time scale of the history of mathematics. The first set of numbers to be used was probably the set of counting numbers $\{1, 2, 3, \ldots\}$. In this chapter we will encounter mathematical ideas that center around the set of counting numbers and the set of nonnegative integers $\{0, 1, 2, \ldots\}$.

7.1 MATHEMATICAL INDUCTION

Goals

Upon completion of this section you should be able to:

1. Understand the principle of mathematical induction.
2. Use the principle of mathematical induction to prove some mathematical statements.

* *

The student is already aware of the fact that one exception to a mathematical rule or formula is all that is necessary to prove it to be false. This means that in order for a statement to be true about a set, it must be true for every element within the set. Whenever a set is infinite, it is, of course, not possible to check the statement for every element of the set. *Mathematical induction* is a method of proof used when we have a statement concerning the infinite set of positive integers. This method is based on the following theorem.

The Principle of Mathematical Induction

If a statement P is true for the integer 1 and if the assumption that P is true for the integer k implies that P is true for $(k + 1)$, then P is true for the set of positive integers.

A theoretical example might make this principle more meaningful.

Suppose there is a step ladder with infinitely many steps. What two things would be necessary for you to know so that you could climb the ladder to any height? A little thought should produce the following two items:

1. You must know that you can at least get on the first step of the ladder.
2. You must know that if you are on any step of the ladder, you can get to the next step.

A little reflection will convince you that these two concepts are not only necessary but also sufficient for you to climb the ladder to any height. Do you see that they represent the statement of the principle of mathematical induction?

We now look at a mathematical example.

EXAMPLE

Prove that the sum of integers 1 through n is given by the formula $\frac{(n^2 + n)}{2}$. In other words, show that

$$1 + 2 + 3 + \cdots + n = \frac{n^2 + n}{2}$$

Step 1 Is the formula true for $n = 1$?

$$\frac{1^2 + 1}{2} = \frac{1 + 1}{2} = 1$$

Thus the formula holds true for $n = 1$.

Step 2 Assume the formula is true for $n = k$.

$$1 + 2 + 3 + \cdots + k = \frac{k^2 + k}{2}$$

Step 3 We must now show that the assumption of Step 2 will imply that the sum of the first $(k + 1)$ integers is $\frac{(k + 1)^2 + (k + 1)}{2}$.
We know from Step 2 that

$$1 + 2 + 3 + \cdots + k = \frac{k^2 + k}{2}$$

Now if we add $(k + 1)$ to each side, we obtain

$$1 + 2 + 3 + \cdots k + (k + 1) = \frac{k^2 + k}{2} + (k + 1)$$

$$= \frac{k^2 + k + 2k + 2}{2}$$

$$= \frac{k^2 + 3k + 2}{2}$$

At this point, we should always reflect on the form we desire and use any necessary algebraic manipulations to obtain it. The sum of the first $(k + 1)$ integers must fit the original formula, $\frac{(n^2 + n)}{2}$, with $n = (k + 1)$. We thus write $k^2 + 3k + 2$ as $k^2 + 2k + 1 + k + 1$ and factor by grouping the first three terms so that we get $(k + 1)^2 + (k + 1)$. We now have the desired result

$$1 + 2 + 3 + \cdots + (k + 1) = \frac{(k + 1)^2 + (k + 1)}{2}$$

which shows that the formula holds for $n = k + 1$. Thus by the principle of mathematical induction, the formula is true for n equal to any integer.

Any and all manipulative skills of algebra can be used to rearrange the right side of Step 3 to obtain the proper form. A great deal of ingenuity is sometimes required.

EXAMPLE

Prove $n < 2^n$ for all positive integers n.

Step 1 Show the statement is true for $n = 1$.
$$1 < 2^1 \quad \text{or} \quad 1 < 2$$

Step 2 Assume the statement is true for $n = k$.
$$k < 2^k$$

Step 3 Show that the assumption of Step 3 implies that $k + 1 < 2^{k+1}$. From Step 2 we have
$$k < 2^k$$

We also know that $1 < 2^k$ so we may add this to the inequality $k < 2^k$ obtaining

SEQUENCES AND SERIES

$$k + 1 < 2^k + 2^k$$
$$k + 1 < 2^k(1 + 1)$$
$$k + 1 < 2^k(2)$$
$$k + 1 < 2^{k+1}$$

Therefore, the statement is true for all n.

Exercise 7.1.1

Prove the following statements by using mathematical induction:

1. $1 + 3 + 5 + \cdots + (2n - 1) = n^2$
2. $2 + 6 + 10 + \cdots + (4n - 2) = 2n^2$
3. $1 + 4 + 7 + \cdots + (3n - 2) = \dfrac{3n^2 - n}{2}$
4. $1^3 + 2^3 + 3^3 + \cdots + n^3 = \dfrac{n^2(n + 1)^2}{4}$
5. $\dfrac{1}{1(2)} + \dfrac{1}{2(3)} + \dfrac{1}{3(4)} + \cdots + \dfrac{1}{n(n + 1)} = \dfrac{n}{n + 1}$
6. $3 + 3^2 + 3^3 + \cdots + 3^n = \dfrac{3(3^n - 1)}{2}$
7. $1(2) + 2(3) + 3(4) + \cdots + n(n + 1) = \dfrac{n(n + 1)(n + 2)}{3}$
8. $1 < x^n$ for all $x > 1$
9. $1 + 4n \leq 5^n$
10. $1 + x + x^2 + \cdots + x^{n-1} = \dfrac{x^n - 1}{x - 1}, \quad x \neq 1$

7.2 SEQUENCES AND SERIES

Goals

Upon completion of this section you should be able to:

1. Define a sequence.
2. Distinguish between a finite and infinite sequence.
3. Define a series.
4. Write a series in sigma notation.

* *

A second topic involving the positive integers is that of sequences and series. A sequence can be defined as any array of numbers such that there is a first, second, third, and so forth. A more formal definition follows.

Definition A *sequence* is a function whose domain is the set of positive integers.

Note that the definition of a function would require one term for each positive integer. Thus $f(1)$ would be the first term, $f(2)$ the second term, etc. Subscripts are generally used to designate the number of a term. Instead of $f(1), f(2), f(3), \ldots, f(n)$, it is generally customary to use $a_1, a_2, a_3, \ldots, a_n$ to designate a sequence with n terms.

Definition A sequence with a finite number of terms is called a *finite sequence,* and a sequence with an infinite number of terms is called an *infinite sequence.*

To avoid any misunderstanding about ordering, we will designate the nth term (a_n) of a sequence by an algebraic formula.

EXAMPLE Write the first four terms and the tenth term of $a_n = \dfrac{n}{n+1}$.

$$a_1 = \frac{1}{1+1} = \frac{1}{2}$$

$$a_2 = \frac{2}{2+1} = \frac{2}{3}$$

$$a_3 = \frac{3}{3+1} = \frac{3}{4}$$

$$a_4 = \frac{4}{4+1} = \frac{4}{5}$$

$$\vdots$$

$$a_{10} = \frac{10}{10+1} = \frac{10}{11}$$

A *recursive* definition of a sequence occurs when we give the first term and then give a method of finding the $(k + 1)$ term from the kth term.

EXAMPLE Find the first four terms of the sequence if $a_1 = 1$ and $a_{k+1} = 3a_k - 1$.

$$a_1 = 1$$
$$a_2 = 3a_1 - 1 = 3(1) - 1 = 2$$
$$a_3 = 3a_2 - 1 = 3(2) - 1 = 5$$
$$a_4 = 3a_3 - 1 = 3(5) - 1 = 14$$

We could, in this manner, find as many terms of this sequence as we wish. However, if we wanted to find the 20th term, we would need to first find the first 19 terms. It is desirable to have a formula that would give us the nth term without having to find all the preceding terms.

For the sequence just given, the formula $a_n = \frac{1}{2}(3^{n-1} + 1)$ is valid for the terms we have obtained. To prove this formula for the entire sequence, we must use mathematical induction.

EXAMPLE If $a_1 = 1$ and $a_{k+1} = 3a_k - 1$, prove $a_n = \frac{1}{2}(3^{n-1} + 1)$ for all positive integers n.

Step 1 Show the formula holds for $n = 1$.

$$a_1 = \frac{1}{2}(3^{1-1} + 1) = \frac{1}{2}(2) = 1$$

Step 2 Assume the statement is true for $n = k$.

$$a_k = \frac{1}{2}(3^{k-1} + 1)$$

Step 3 Show that the $(k + 1)$ term is $\frac{1}{2}\left[3^{(k+1)-1} + 1\right]$.

We know from the recursive definition that the $(k + 1)$ term is $3a_k - 1$. Therefore using the value for a_k assumed in Step 2, the $(k + 1)$ term is

$$3a_k - 1 = 3\left[\frac{1}{2}(3^{k-1} + 1)\right] - 1$$
$$= \frac{1}{2}(3^k + 3) - 1$$

7.2 SEQUENCES AND SERIES

$$= \frac{1}{2}(3^k + 3 - 2)$$

$$= \frac{1}{2}(3^k + 1)$$

$$= \frac{1}{2}\left[3^{(k+1)-1} + 1\right]$$

This last expression is our formula with $n = (k + 1)$, and hence by mathematical induction the formula holds for all positive integers n.

Definition A *series* is the indicated sum of a sequence.

EXAMPLE If the sequence is 1, 3, 5, 7, 9, then the series is

$$1 + 3 + 5 + 7 + 9$$

A special notation is sometimes used to indicate a series. It is especially useful when we have an expression for the general term of a sequence. The upper case Greek letter sigma (Σ) is used to indicate a sum.

EXAMPLE Find $\sum_{n=1}^{5}(2n - 1)$.

The notation is read as "Find the sum of the terms $2n - 1$, as n takes on successive integral values from 1 to 5." We get

$$[2(1) - 1] + [2(2) - 1] + [2(3) - 1] + [2(4) - 1] + [2(5) - 1]$$
$$= 1 + 3 + 5 + 7 + 9$$
$$= 25$$

Therefore $\sum_{n=1}^{5}(2n - 1) = 25$.

Note that any letter may be used to indicate the variable.

EXAMPLES

$$\sum_{i=1}^{8} i^2 = 1^2 + 2^2 + 3^2 + \cdots + 8^2$$

$$\sum_{k=1}^{4} (k+1)^2 = 2^2 + 3^2 + 4^2 + 5^2$$

Exercise 7.2.1

Write the first four terms and the tenth term of the sequence whose nth term is defined as follows:

1. $a_n = 3n - 2$
2. $a_n = \dfrac{n}{2} + 1$
3. $a_n = n^2 + 3$
4. $a_n = 2n^2 - 5n + 1$
5. $a_n = \dfrac{\log n}{2n}$

Find the first four terms of each of the sequences defined recursively as follows:

6. $a_1 = 2,\ a_{k+1} = 3a_k - 4$
7. $a_1 = -4,\ a_{k+1} = 5a_k + 1$
8. $a_1 = \dfrac{1}{2},\ a_{k+1} = \dfrac{a_k}{2}$
9. $a_1 = 1,\ a_{k+1} = (a_k)^2 + 3$
10. $a_1 = 2,\ a_{k+1} = ka_k - 1$

Find the value of each of the following:

11. $\displaystyle\sum_{n=1}^{4} 5n$
12. $\displaystyle\sum_{k=3}^{5} \dfrac{1}{2}k$
13. $\displaystyle\sum_{i=0}^{4} (2^i + 1)$
14. $\displaystyle\sum_{k=2}^{5} \dfrac{k}{k-1}$
15. $\displaystyle\sum_{j=1}^{5} \dfrac{j+2}{2^{j-1}}$

Write each of the following series in sigma notation:

16. $1 + 2 + 3 + 4 + 5$
17. $4 + 9 + 16 + 25$
18. $\dfrac{1}{3} + \dfrac{1}{9} + \dfrac{1}{27} + \dfrac{1}{81}$
19. $1 - 1 + 1 - 1 + 1 - 1$
20. $1 + 6 + 11 + 16 + 21$

7.2 SEQUENCES AND SERIES

7.3 ARITHMETIC SEQUENCES

Goals Upon completion of this section you should be able to:
1. Define an arithmetic sequence.
2. Find the nth term of an arithmetic sequence.
3. Find the sum of the first n terms of an arithmetic sequence.

* *

Even though the definition of a sequence allows any list of numbers to qualify as a sequence, we have found that generally the terms are related in some way to preceding terms or to the counting numbers of the terms. In this section we will investigate one special type of sequence known as an arithmetic sequence.

Definition An *arithmetic sequence* is a sequence in which each term is obtained by adding a constant (we will designate the constant by d) to the preceding term.

Note that this definition is recursive since $a_{k+1} = a_k + d$. Given the first term and d, we could write the first n terms.

EXAMPLE If the first term of an arithmetic sequence is 7 and $d = 4$, the first five terms are

$$7, 11, 15, 19, 23$$

If we designate the first term of an arithmetic sequence by a and the constant difference by d, then the sequence is

$$a, a + d, (a + d) + d, [(a + d) + d] + d, \text{etc.}$$

or
$$a, a + d, a + 2d, a + 3d, \text{etc.}$$

An inspection of this sequence leads us to a formula for the nth term of an arithmetic sequence.

Theorem If a is the first term and d is the constant difference in an arithmetic sequence, then the nth term is given by $a_n = a + (n - 1)d$.

A formal proof of this theorem could be given using mathematical induction.

EXAMPLE

Find the tenth term of an arithmetic sequence where $a = 3$ and $d = 7$.

$$a_n = a + (n-1)d$$
$$a_{10} = 3 + (10-1)7$$
$$= 66$$

Theorem

The sum S_n of the first n terms of an arithmetic series is given by the formula $S_n = \dfrac{n}{2}(a_1 + a_n)$ where a_1 is the first term and a_n is the nth term.

Proof By summing the sequence of n terms having the first term as a_1 and the constant difference of d, we have

$$S_n = a_1 + (a_1 + d) + (a_1 + 2d) + \cdots + [a_1 + (n-1)d]$$

If we now write the series in reverse order, we obtain

$$S_n = a_n + (a_n - d) + (a_n - 2d) + \cdots + [a_n - (n-1)d]$$

Adding the two expressions gives

$$2S_n = (a_1 + a_n) + (a_1 + a_n) + (a_1 + a_n) + \cdots + (a_1 + a_n)$$

which can be condensed to

$$2S_n = n(a_1 + a_n)$$

or

$$S_n = \frac{n}{2}(a_1 + a_n)$$

Since the nth term is $a_n = a_1 + (n-1)d$, we can rewrite the formula as

$$S_n = \frac{n}{2}[2a_1 + (n-1)d]$$

EXAMPLE

If the first term of an arithmetic sequence is 7 and the constant difference is 4, find the following:

1. The third term.
2. The fourteenth term.
3. The sum of the first ten terms.

    1. $a_3 = 7 + (3-1)4$

           $= 7 + (2)4$

           $= 15$

2. $a_{14} = 7 + (14 - 1)4$
   $= 7 + (13)4$
   $= 59$

3. $S_{10} = \dfrac{10}{2}[2(7) + (9)(4)]$
   $= 5(14 + 36)$
   $= 250$

Exercise 7.3.1

1. Write the first six terms of the arithmetic sequence where $a = 3$ and $d = 5$.
2. Write the first seven terms of the arithmetic sequence where $a = 8$ and $d = -3$.

Find the indicated term in each of the arithmetic sequences in Problems 3 to 8.

3. $a = 2, d = 5$: 11th term
4. $a = 6, d = \dfrac{1}{2}$: 13th term
5. $a = 3, d = -2$: 6th term
6. $a = \dfrac{1}{2}, d = \dfrac{1}{3}$: 20th term
7. $a = -3, d = 4$: 50th term
8. $a = -4, d = -\dfrac{1}{3}$: 8th term

Find the sum of each of the arithmetic sequences in Problems 9 to 13.

9. $a = 2, d = 4, n = 10$
10. $a = 3, d = \dfrac{1}{2}, n = 9$
11. $a = 10, d = -4, n = 8$
12. $a = -\dfrac{1}{2}, d = 5, n = 10$
13. $a = 7, d = -\dfrac{3}{2}, n = 12$

Problems 14 to 20 all refer to arithmetic sequences.

14. If $a_1 = 7$, $a_n = 91$, and $S_n = 343$, find n.
15. If $a_{16} = -20$ and $S_{16} = -80$, find a_1.
16. If $a_n = 11$, $d = 2$, and $S_n = 32$, find n.
17. If $a_1 = 1$, $d = \frac{1}{2}$, and $S_n = 27$, find a_n.
18. Find the sum of the first 100 positive integers.
19. For what value of k is the sequence $2k + 4$, $3k - 7$, $k + 12$ an arithmetic sequence?
20. If the numbers a, b, c, d, e form an arithmetic sequence, then b, c, and d are called *arithmetic means* between a and e. Find three arithmetic means between -12 and 60.
21. A sequence of numbers is called a *harmonic progression* if their reciprocals form an arithmetic sequence. The sequence $1, \frac{1}{3}, \frac{1}{6}, \frac{1}{9}, \frac{1}{12}$ is a harmonic progression since their reciprocals $1, 3, 6, 9, 12$ form an arithmetic sequence. Is the sequence $\frac{6}{7}, \frac{3}{5}, \frac{6}{13}, \frac{3}{8}$ a harmonic progression?
22. The *harmonic mean* of two numbers is the reciprocal of the arithmetic mean of the reciprocals of the two numbers. Find the harmonic mean of $\frac{1}{5}$ and $\frac{1}{8}$.
23. A person borrows $10,000 for 10 years and agrees to pay back $1,000 principal each year plus simple interest at 8% on all principal left unpaid during the year. How much interest is paid over the 10-year period?

7.4 GEOMETRIC SEQUENCES

Goals

Upon completion of this section you should be able to:
1. Define a geometric sequence.
2. Find the nth term of a geometric sequence.
3. Find the sum of the first n terms of a geometric sequence.
4. Find the sum of an infinite geometric sequence.

* *

Another sequence of special interest is the geometric sequence.

Definition A *geometric sequence* is a sequence in which each term is obtained by multiplying the preceding term by a constant ratio (which we will denote by r).

Note that this definition is a recursive definition. If we denote the first term as a_1 and the constant ratio as r, we obtain the expansion of the geometric sequence as

$$a_1, a_1 r, a_1 r^2, a_1 r^3, \ldots, a_1 r^{n-1}$$

Theorem **The formula for the nth term of a geometric sequence having a first term a_1 and a constant ratio r is $a_n = a_1 r^{n-1}$.**

EXAMPLE Write the first four terms and the seventh term of a geometric sequence if the first term is 3 and the constant ratio is 2.

The first four terms are 3, 6, 12, 24. The seventh term is

$$\begin{aligned} a_7 &= 3(2)^6 \\ &= 3(64) \\ &= 192 \end{aligned}$$

Theorem **The sum of the first n terms of a geometric sequence is**

$$S_n = \frac{a_1 - a_1 r^n}{1 - r}$$

Proof By summing the terms of the sequence, we have

$$S_n = a_1 + a_1 r + a_1 r^2 + \cdots + a_1 r^{n-1}$$

If we multiply both sides of the equality by r, we obtain

$$r S_n = a_1 r + a_1 r^2 + a_1 r^3 + \cdots + a_1 r^{n-1} + a_1 r^n$$

Subtracting the last expression from the first gives

$$S_n - r S_n = a_1 - a_1 r^n$$
$$S_n (1 - r) = a_1 - a_1 r^n$$
$$S_n = \frac{a_1 - a_1 r^n}{1 - r}$$

Another form of this formula may be obtained by observing that

$$a_n = a_1 r^{n-1}$$

Multiplying by r gives
$$a_n r = a_1 r^n$$
If we replace $a_1 r^n$ by $a_n r$ in the preceding formula, we obtain
$$S_n = \frac{a_1 - a_n r}{1 - r}$$

EXAMPLE

Find the sum of the first six terms of the geometric sequence in which $a_1 = \frac{2}{3}$ and $r = \frac{1}{2}$.

$$S_6 = \frac{a_1 - a_6 r}{1 - r}$$

$$= \frac{\frac{2}{3} - \left(\frac{1}{48}\right)\left(\frac{1}{2}\right)}{1 - \frac{1}{2}}$$

$$= \frac{\frac{2}{3} - \frac{1}{96}}{\frac{1}{2}}$$

$$= \frac{63}{48}$$

Definition

If the numbers $a_1, a_2, a_3, \ldots, a_n$ form a geometric sequence, the numbers $a_2 \ldots a_{n-1}$ are called *geometric means* between a_1 and a_n.

EXAMPLE

Insert three geometric means between 3 and 48. We are given that $a_1 = 3$ and $a_5 = 48$. We know that
$$a_5 = a_1 r^4$$
so
$$48 = 3r^4$$
$$r^4 = 16$$
$$r = \pm 2$$

Hence, the sequence is
$$3, 6, 12, 24, 48$$
or
$$3, -6, 12, -24, 48$$

Exercise 7.4.1

1. Write the first five terms of a geometric sequence where the first term is 2 and the constant ratio is 3.
2. Write the first four terms of a geometric sequence where the first term is 10 and the constant ratio is −2.

Find the indicated term in each of the geometric sequences in Problems 3 to 6.

3. $a_1 = 3, r = 5$: 6th term
4. $a_1 = 8, r = \dfrac{3}{2}$: 5th term
5. $a_1 = -3, r = -2$: 8th term
6. $a_1 = 4, r = -\dfrac{1}{2}$: 6th term

Each of the Problems 7 to 10 refers to geometric sequences.

7. If $a_1 = 1, r = -2$, and $S_n = -21$, find n.
8. If $a_7 = 192$ and $r = 2$, find S_7.
9. If $a_1 = 4$ and $S_4 = -80$, find r.
10. If $r = -3$ and $S_7 = 2{,}188$, find a_7.
11. If log 3 is the first term of a geometric sequence and log 9 is the second term, find the third and fourth terms.
12. For what value of k is the sequence $k - 2, k - 6, 2k + 3$ a geometric sequence?
13. Insert four geometric means between 3 and 96.
14. Insert three geometric means between 16 and 625.
15. If you saved 1 cent the first day and added 2 cents the second day, 4 cents the third day, etc., how many days would it take to save 1 million dollars?

The formula for S_n gives us the sum of the first n terms in a geometric sequence. The question arises "Can we find the sum of the terms in an infinite geometric sequence?" The answer is "sometimes," and the determining factor is the size of the constant ratio r.

A study of limits is necessary to prove the following theorem, and such a study is beyond the scope of this text. We therefore state the theorem without proof and leave as an exercise an intuitive development of the formula.

Theorem

The sum S of an infinite geometric sequence $a_1, a_1 r, a_1 r^2, \ldots$ with $|r| < 1$ is given by the formula

$$S = \frac{a_1}{1 - r}$$

EXAMPLE

Find the sum of the infinite geometric sequence

$$\frac{2}{3}, \frac{2}{9}, \frac{2}{27}, \ldots$$

We see that $a_1 = \frac{2}{3}$ and $r = \frac{1}{3}$. Since $|r| < 1$ we may write

$$S = \frac{a_1}{1 - r}$$

$$= \frac{\frac{2}{3}}{1 - \frac{1}{3}}$$

$$= 1$$

Exercise 7.4.2

Find the sum, if it exists, of each infinite geometric series in Problems 1 to 7.

1. $4 + 2 + 1 + \frac{1}{2} + \cdots$
2. $1 - \frac{1}{2} + \frac{1}{4} - \frac{1}{8} + \cdots$
3. $\sqrt{3} + 3 + \sqrt{27} + 9 + \cdots$
4. $15 - 5 + \frac{5}{3} - \frac{5}{9} + \cdots$
5. $4 + 2\sqrt{2} + 2 + \sqrt{2} + \cdots$
6. $1 + 0.5 + 0.25 + 0.125 + \cdots$
7. $\sum_{i=1}^{\infty} \left(\frac{2}{3}\right)^i$

8. Given the decimal number $0.3333\cdots$
   a. Write this number as an infinite geometric series.
   b. Find a_1 and r.

7.4 GEOMETRIC SEQUENCES

c. Find the sum of the series.

9. Repeat the directions in Problem 8 for the following decimals:

   a. 0.363636 · · ·

   b. 0.345345 · · ·

10. Discuss what happens to the formula $S_n = \dfrac{a_1 - a_1 r^n}{1 - r}$ if $|r| < 1$ and n becomes very large.

11. A rubber ball is dropped from a height of 12 meters. Each time it strikes the ground it rebounds three-fourths of the height from which it fell. What is the approximate total distance traveled by the ball?

CHAPTER REVIEW

Use mathematical induction to prove the statements in Problems 1 to 6.

1. $\dfrac{1}{2}(1) + \dfrac{1}{2}(2) + \dfrac{1}{2}(3) + \cdots + \dfrac{1}{2}(n) = \dfrac{n^2 + n}{4}$

2. $5 + 9 + 13 + \cdots + (4n + 1) = 2n^2 + 3n$

3. $2 + 2^2 + 2^3 + \cdots + 2^n = 2^{n+1} - 2$

4. $1 + 3 + 6 + \cdots + \dfrac{n(n+1)}{2} = \dfrac{n(n+1)(n+2)}{6}$

5. $1^2 + 3^2 + 5^2 + \cdots + (2n-1)^2 = \dfrac{n(4n^2 - 1)}{3}$

6. $\dfrac{1}{1(3)} + \dfrac{1}{3(5)} + \dfrac{1}{5(7)} + \cdots + \dfrac{1}{(2n-1)(2n+1)} = \dfrac{n}{2n+1}$

7. Write the first four terms and the eighth term of the sequence whose nth term is given as $a_n = n^3 - 4$.

8. Write the first and the 112th term of the sequence whose nth term is $a_n = (-1)^n + 5$.

9. Find the first four terms of the sequence that is defined recursively as $a_1 = 1$, $a_{k+1} = k(a_k + 1)$.

10. Evaluate $\sum\limits_{i=1}^{5} 2^{i-1}$.

11. Express in expanded form: $\sum\limits_{i=1}^{4} \dfrac{x^{i+2}}{4i}$

12. Express the series $2 + 5 + 10 + 17 + 26$ in sigma notation.

13. Express the series $1 + 5 + 9 + 13 + 17 + 21$ in sigma notation.

14. In an arithmetic sequence, if $a = -8$ and $d = 5$, find the 18th term.

15. In an arithmetic sequence, if $a = 8$ and $d = -\frac{1}{2}$, find the 29th term.

16. In an arithmetic sequence, if $a = x + 2y$ and $d = 2x + y$, find the 12th term.

17. Find the sum of the first 100 positive odd integers.

18. If the sum of the first nine terms of an arithmetic sequence where $d = -3$ is 468, find the first term.

19. An arithmetic sequence having ten terms has a first term of 18 and a sum of 45. Find the seventh term.

20. For what value of k is the sequence $k - 3, k + 5, 2k - 1$ an arithmetic sequence?

21. Write the first four terms of a geometric sequence where the first term is $\sqrt{2}$ and the constant ratio is $\sqrt{2}$.

22. Find the eighth term of a geometric sequence where the first term is 6 and the constant ratio is 2.

23. Find the fifth term of a geometric sequence where the first term is -8 and the constant ratio is $-\frac{1}{2}$.

24. Find the sum of the first six terms of a geometric sequence where $a_1 = 16$ and $r = -\frac{1}{4}$.

25. For what values of k is the sequence $3k + 4, k - 2, 5k + 1$ a geometric sequence?

26. Insert three geometric means between 81 and 16.

27. Using an infinite geometric series, find the common fraction that is equal to $0.6666\cdots$.

PRACTICE TEST

1. Use mathematical induction to prove $2 + 4 + 6 + \cdots + 2n = n(n + 1)$.

2. Write the first four terms of the sequence whose nth term is defined as $a_n = 3n^2 - n - 2$.

3. Evaluate: $\sum_{i=2}^{7} (2i - 1)$.

4. Express the series $-1 + 2 + 7 + 14$ in sigma notation.

5. If the first term of an arithmetic sequence is $\frac{1}{3}$ and the constant difference is $\frac{1}{2}$, find the ninth term.

6. The 14th term of an arithmetic sequence is 81 and the 20th term is 117. Find the sum of the first four terms.

7. In an arithmetic sequence, if $a_1 = 3$, $a_n = 27$, and $S_n = 150$, find n.

8. The first term of a geometric sequence is 625 and the last term is 1. If the sum of the terms is 521, find the fourth term.

9. Insert three geometric means between $\frac{2}{3}$ and 54.

10. Find the sum of the infinite geometric series $6 - 4 + \frac{8}{3} - \frac{16}{9} + \cdots$.

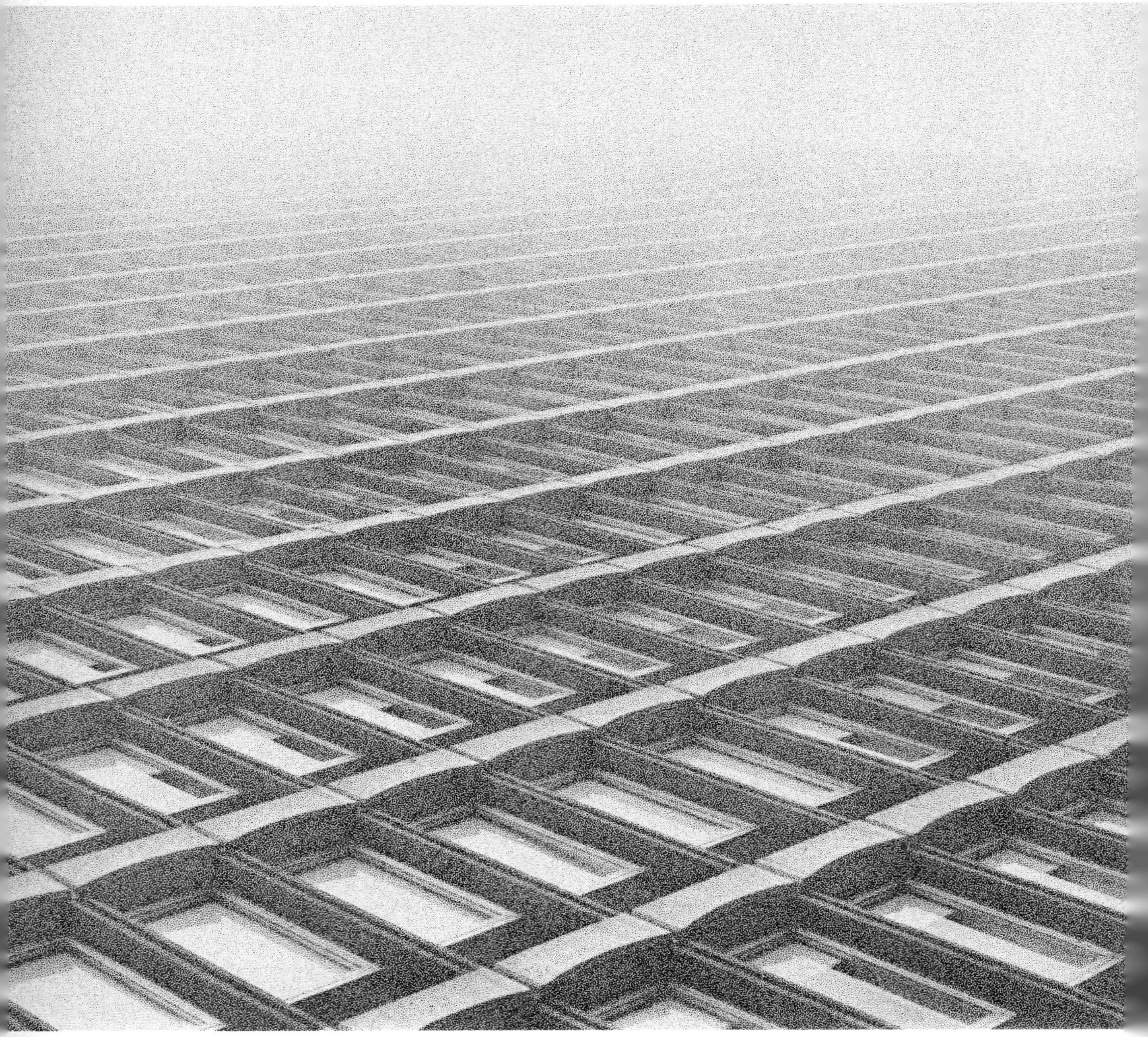

The Binomial Theorem
and an Introduction to Probability

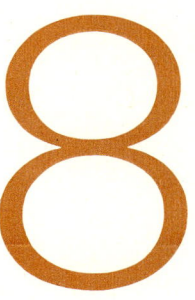

8

8

The study of statistics has become an increasingly important focus for mathematics in recent years. Much of modern industry, for example, depends heavily on the application of statistics in developing quality control. The topics in this chapter are, in part, background material for a study of statistics. The introductory sections on permutations and combinations provide information used in the binomial theorem and probability as well as in later courses in statistics.

8.1 PERMUTATIONS

Goals

Upon completion of this section you should be able to:

1. Define a permutation.
2. Use the fundamental counting principle to determine permutations.
3. Use the notation $_nP_r$ and factorial notation.

* *

If a succession of acts can occur in more than one way, or if a set of objects can be arranged in more than one way, there is often a need to know the total number of ways these events can occur. The topic of *permutations* deals with this question.

Consider the following example:

> A manufacturer of automobiles has four basic body designs. With each body design he can use any of three trim units and any of two engines. How many different automobiles can he produce?

It is easy to see that if with each of four body designs he can use any of three trims, then there are (4)(3) or 12 different arrangements of body and trim. With each of these 12 arrangements he can use two engines giving (12)(2) or 24 different automobiles.

This leads to the following *fundamental counting principle*.

250   THE BINOMIAL THEOREM AND AN INTRODUCTION TO PROBABILITY

Theorem If an act can occur in *r* ways and, after it has occurred, a second act can occur in *s* ways, and thereafter a third act can occur in *t* ways, and so forth, until the last act can occur in *k* ways, then the successive acts can occur in (*r*)(*s*)(*t*) ... (*k*) ways.

Definition The arrangement of a group of objects in a definite order is called a *permutation* of the objects.

EXAMPLE How many permutations are possible with the three letters *a*, *b*, and *c*? We can solve the problem by listing the arrangements.

a, b, c	*b, a, c*	*c, a, b*
a, c, b	*b, c, a*	*c, b, a*

We see that there are six permutations.

If the number of objects is large, such a listing would be extremely difficult. We will apply the fundamental counting principle to the problem in the following manner. Let three blanks represent the three positions of the letters.

___ ___ ___

First we can place a letter in the first position. We have three choices.

_3_ ___ ___

Now we have only two choices for the second position since one of the letters has already been used.

_3_ _2_ ___

We now have only one choice for the third letter.

_3_ _2_ _1_

By the fundamental counting principle, this gives us (3)(2)(1) = 6 arrangements. If we wished to permute four objects, we would have (4)(3)(2)(1) = 24 permutations. In general, the number of ways in which *n* objects could be permuted is

$$n(n-1)(n-2) \ldots (3)(2)(1)$$

In order to provide consistency for future formulas involving factorial notation, we make the following definitions.

8.1 PERMUTATIONS

Definition   If n is a positive integer, then the symbol $n!$ (read "n factorial") is defined as $n(n-1)(n-2) \ldots (3)(2)(1)$.

For example, $5! = (5)(4)(3)(2)(1) = 120$.

Definition $0! = 1$

Theorem   **The number of permutations of n things taken all at a time is equal to $n!$.**

The permutation of n things taken less than n at a time would of course be less than $n!$.

Consider the number of ways the five letters a, b, c, d, e can be arranged using only three at a time. The first of the three positions can be filled in any one of five ways giving

$$\underline{5} \quad \underline{} \quad \underline{}$$

The second can be filled in four ways and the third in three ways. We have

$$\underline{5} \quad \underline{4} \quad \underline{3}$$

or $(5)(4)(3) = 60$ permutations of five things taken three at a time.

To generalize, if we have n things and use only r at a time, we would need r positions. The first position could be filled in n ways, the second in $(n-1)$ ways, and so forth. The number of objects that would remain for the rth or last position would be $n-(r-1)$.

Theorem   **The number of permutations of n things taken r at a time is $n(n-1)(n-2) \ldots [n-(r-1)]$.**

Definition   The notation for the permutation of n things taken r at a time is ${}_nP_r$.

The formula for ${}_nP_r$ can be written in factorial notation if we note that

$${}_nP_r = n(n-1)(n-2) \ldots [n-(r-1)]$$

$$= n(n-1)(n-2) \ldots [n-(r-1)] \cdot \frac{(n-r)!}{(n-r)!}$$

$$= \frac{n!}{(n-r)!}$$

252   THE BINOMIAL THEOREM AND AN INTRODUCTION TO PROBABILITY

We will generally use the formula as

$$_nP_r = \frac{n!}{(n-r)!}$$

EXAMPLE

In how many ways can twelve objects be arranged if we use three at a time?

$$_{12}P_9 = \frac{12!}{(12-3)!}$$

$$= \frac{12!}{9!}$$

$$= \frac{(12)(11)(10)(9)(8)(7)(6)(5)(4)(3)(2)(1)}{(9)(8)(7)(6)(5)(4)(3)(2)(1)}$$

$$= (12)(11)(10)$$

$$= 1,320$$

We could have shortened the work somewhat by writing $\frac{12!}{9!}$ as $\frac{(12)(11)(10)(9!)}{9!}$ and then reducing to $(12)(11)(10) = 1,320$.

Exercise 8.1.1

A calculator may be useful in some of the following problems. Evaluate the expressions in Problems 1 to 10.

1. $5!$
2. $10!$
3. $(4!)(3!)$
4. $\frac{20!}{18!}$
5. $_6P_4$
6. $_8P_3$
7. $_{10}P_5$
8. $_{12}P_4$
9. $_5P_5$
10. $_7P_7$

11. In how many ways can the letters in the word "triangle" be arranged?

12. How many five-digit numerals can be formed from the digits $\{2, 3, 4, 5, 6\}$ if no digit is repeated?

13. How many three-digit numerals can be formed from the digits $\{1, 2, 3, 4, 5, 6\}$ if no digit is repeated?

14. In how many ways can six people be arranged in a straight line?

15. In how many ways can the letters of the alphabet be arranged in a sequence of five letters if no letter is repeated?

16. A disc jockey wishes to play eleven records on a radio program. How many different orders of playing them are there?

8.1 PERMUTATIONS

17. Twelve horses are entered in a race. In how many ways could they finish first, second, and third?

18. In how many ways can six people be arranged in a circle?

19. How many even four-digit numerals can be formed from the digits $\{1, 2, 3, 4, 5\}$ if digits are allowed to repeat?

20. How many different automobile license plates can be made if each contains three letters followed by three numbers and

   a. if no letter or number is repeated?

   b. if the letters and numbers may be repeated?

8.2 COMBINATIONS

Goals

Upon completion of this section you should be able to:

1. Define a combination.
2. Understand the notation and use the formula for $\binom{n}{r}$.

* *

In many instances we are interested in a collection of items but the order in which they are arranged is not important. In such cases we are dealing with *combinations.*

Definition

A set of elements in which order is not considered is called a *combination.*

The context of a problem will distinguish between permutations and combinations. For instance, if we ask "How many four-digit numbers can be written using the digits $\{1, 2, 3, 4, 5, 6\}$ without repeating any one digit in a given number?", we have a permutation problem since the different order of the digits would give a different number. If we ask "How many different four-member committees can be selected from six people?", we have a combination problem since the order of selection would not make a different committee.

Definition

The notation for the combination of n things taken r at a time is $\binom{n}{r}$.

Theorem

The formula for the combination of n things taken r at a time is

$$\binom{n}{r} = \frac{n!}{(n-r)!r!}$$

Proof Suppose we know the number of combinations of n things taken r at a time. Each of these combinations would contain r elements which could be permuted in $r!$ ways, so

$$\binom{n}{r} r! = {}_nP_r$$

This shows the relationship between permutations of n things taken r at a time and the combinations of n things taken r at a time. Thus

$$\binom{n}{r} r! = {}_nP_r$$

$$\binom{n}{r} r! = \frac{n!}{(n-r)!}$$

$$\binom{n}{r} = \frac{n!}{(n-r)!r!}$$

EXAMPLE

How many four-member committees can be formed from a group of nine people?

$$\binom{9}{4} = \frac{9!}{(9-4)!4!}$$

$$= \frac{9!}{5!\,4!}$$

$$= \frac{(9)(8)(7)(6)(5)(4)(3)(2)(1)}{(5)(4)(3)(2)(1)(4)(3)(2)(1)}$$

$$= 126$$

Again, we could have shortened the solution by writing

$$\frac{9!}{5!4!} \quad \text{as} \quad \frac{(9)(8)(7)(6)(5!)}{5!(4)(3)(2)(1)}$$

and then have reduced the answer.

Exercise 8.2.1

A calculator may be useful in some of the following problems. Evaluate the expressions in Problems 1 to 4.

1. $\binom{10}{3}$ 2. $\binom{10}{7}$ 3. $\binom{12}{8}$ 4. $\binom{12}{4}$

8.2 COMBINATIONS

5. Show that $\binom{n}{r} = \binom{n}{n-r}$.

6. How many three-member committees can be formed from a group of twelve people?

7. A man has a penny, a nickel, a dime, a quarter, and a half-dollar in his pocket from which he randomly draws three coins. How many ways can he do this?

8. A student must select five classes from a schedule of eight. How many ways are there of making the selection?

9. A woman has found eight art prints that she likes but can only buy three. How many ways could she select them?

10. A district office of a national company is requested to transfer three of its salespersons to the head office. If they have ten salespersons to choose from, in how many ways could they make the selection?

11. How many ways can a person select five gifts from a display of eleven gifts?

12. A class of twenty-five students is asked to subdivide into groups of five members each. In how many ways could this be done?

13. Five cards are drawn at random from a deck of 52 cards. How many different combinations are possible?

14. In a test, a student is to choose ten questions from a set of twelve.

    a. In how many ways may the questions be selected?

    b. If the student is required to answer Questions 1 and 2, in how many ways may the ten questions be selected?

15. How many committees of any number can be formed from a group of ten people?

8.3 THE BINOMIAL THEOREM

Goals

Upon completion of this section you should be able to:

1. Expand powers of binomials using the binomial theorem.
2. Find any term of a binomial expansion without expanding the binomial.
3. Find powers and roots of numbers by using the binomial theorem.

* *

Raising a binomial to a power occurs in many phases of mathematics, and the coefficients of this expansion have a special place in the study of statistics. In this section we will discuss and use the theorem for the binomial expansion.

Note the following expansions of $x + y$:

$$(x + y)^1 = x + y$$
$$(x + y)^2 = x^2 + 2xy + y^2$$
$$(x + y)^3 = x^3 + 3x^2y + 3xy^2 + y^3$$
$$(x + y)^4 = x^4 + 4x^3y + 6x^2y^2 + 4xy^3 + y^4$$

Several patterns are seen.

1. For $(x + y)^n$, there are $(n + 1)$ terms.
2. The exponents of x descend from n to zero.
3. The exponents of y ascend from zero to n.
4. The numerical coefficients are symmetrical with respect to the "middle" of the expansion.

If the pattern persists, $(x + y)^{10}$ would have eleven terms, the exponents of x would start at x^{10} in the first term and decrease to x^0 in the eleventh term, while the exponents of y would start with y^0 in the first term and increase to y^{10} in the eleventh term. The numerical coefficients of the first and eleventh terms, second and tenth, third and ninth, etc., would be the same.

It is obvious that finding the value of the numerical coefficients is the only part of a binomial expansion that involves any degree of difficulty.

Binomial Theorem

For any natural number n

$$(a + b)^n = \sum_{r=0}^{n} \binom{n}{r} a^{n-r} b^r$$

The proof is by mathematical induction and is a rather long exercise in algebraic manipulation. We will, therefore, leave this proof to the more enterprising student and proceed to use it for expanding binomials.

EXAMPLE

Expand $(x + y)^5$.

$$(x + y)^5 = \sum_{r=0}^{5} \binom{5}{r} x^{5-r} y^r$$

8.3 THE BINOMIAL THEOREM

$$= \binom{5}{0} x^5 + \binom{5}{1} x^4 y + \binom{5}{2} x^3 y^2 + \binom{5}{3} x^2 y^3$$
$$+ \binom{5}{4} xy^4 + \binom{5}{5} y^5$$
$$= x^5 + 5x^4 y + 10x^3 y^2 + 10x^2 y^3 + 5xy^4 + y^5$$

EXAMPLE

Find the eighth term of the expansion of $(x + y)^{12}$. Note that in the eighth term, $r = 7$. Thus the eighth term is

$$\binom{12}{7} x^5 y^7 = 792 x^5 y^7$$

EXAMPLE

Find the third term of $(2x + 3)^8$.

$$(2x + 3)^8 = [(2x) + 3]^8$$

Note that in the third term, $r = 2$. Thus the third term is

$$\binom{8}{2} (2x)^6 (3)^2 = 28(64x^6)(9)$$
$$= 16{,}128 x^6$$

EXAMPLE

Find the value of $(1.01)^7$ to three decimal places using the binomial expansion. We may write $(1.01)^7$ as $(1 + 0.01)^7$. Then

$$(1 + 0.01)^7 = (1)^7 + 7(1)^6 (0.01) + 21(1)^5 (0.01)^2$$
$$+ 35(1)^4 (0.01)^3 + \cdots$$
$$= 1 + 0.07 + 0.0021 + 0.000035 + \cdots$$
$$\approx 1.072$$

Notice that we only needed to use the first three terms of the expansion rather than all eight terms to acquire three-decimal accuracy. If we were asked for the value to more decimal places, we would simply find more terms of the expansion.

An alternate method of finding the coefficients of the terms of the binomial expansion is as follows:

1. The coefficient of the first term is 1.
2. The coefficient of the rth term can be found by multiplying the coefficient of the $(r - 1)$ term by the exponent of x and dividing this quantity by the number of the $(r - 1)$ term.

EXAMPLE Expand $(a + b)^6$.

First note the pattern of the exponents of a and b discussed earlier and see that the first term is a^6. Next multiply 6 (the exponent of a) by 1 (the coefficient of a^6) and divide this quantity by 1 (the number of the term), giving the second term to be $6a^5b$. Now multiply 5 (the exponent of a) by 6 (the coefficient of the second term) and divide this product by 2 (the number of the term) giving $15a^4b^2$ as the third term. Proceeding this way gives

$$(a + b)^6 = a^6 + 6a^5b + 15a^4b^2 + 20a^3b^3 + 15a^2b^4 + 6ab^5 + b^6$$

The validity of this method lies in the fact that $\binom{n}{r}$ [the coefficient of the $(r + 1)$ term] is equal to $\binom{n}{r-1} \cdot \frac{[n - (r - 1)]}{r}$, which is the coefficient of the rth term multiplied by the exponent of a and divided by the number of the term. Proof that

$$\binom{n}{r} = \binom{n}{r-1} \cdot \frac{[n - (r - 1)]}{r}$$

is left as an exercise.

The weakness of this method of expansion lies in the fact that in order to find the eighth term of $(x + y)^{12}$, we would first need to find the first seven terms.

The strength of this method lies in the fact that the arithmetic is easier than expanding by the binomial theorem.

A third method of determining the coefficients of the terms of an expansion of a binomial uses an array of numbers called *Pascal's triangle*. The pattern of this array is indicated below.

$(a + b)^0$						1						
$(a + b)^1$					1		1					
$(a + b)^2$				1		2		1				
$(a + b)^3$			1		3		3		1			
$(a + b)^4$		1		4		6		4		1		
$(a + b)^5$	1		5		10		10		5		1	
	—	—	—	—	—	—	—	—	—	—	—	—

8.3 THE BINOMIAL THEOREM

This pattern may be continued indefinitely. Notice that each row in the array begins and ends with a 1. The second number in any row is obtained by adding the first and second numbers in the preceding row. The third number is obtained by adding the second and third numbers in the preceding row, etc.

The numbers in the nth row of the triangle are the coefficients of the binomial expansion of $(a + b)^{n-1}$. Notice from the triangle, for example, that the coefficients of the expansion $(a + b)^4$ are 1, 4, 6, 4, 1. Thus,

$$(a + b)^4 = a^4 + 4a^3b + 6a^2b^2 + 4ab^3 + b^4$$

EXAMPLE

Using Pascal's triangle, expand $(a + b)^7$.

$(a + b)^0$					1				
$(a + b)^1$				1		1			
$(a + b)^2$				1	2	1			
$(a + b)^3$			1	3		3	1		
$(a + b)^4$			1	4	6	4	1		
$(a + b)^5$		1	5	10		10	5	1	
$(a + b)^6$	1	6	15		20		15	6	1
$(a + b)^7$	1	7	21	35		35	21	7	1

Thus,

$$(a + b)^7 = a^7 + 7a^6b + 21a^5b^2 + 35a^4b^3 + 35a^3b^4 + 21a^2b^5 + 7ab^6 + b^7$$

If we make the assumption that the binomial theorem is valid for rational values of n, we see that the expansion becomes an infinite series in some cases. This can be very useful in creating series that evaluate roots.

EXAMPLE

Use the binomial expansion to find $\sqrt{29}$ correct to three decimal places.

$$\sqrt{29} = (29)^{\frac{1}{2}} = (25 + 4)^{\frac{1}{2}}$$

$$= (25)^{\frac{1}{2}} + \frac{1}{2}(25)^{-\frac{1}{2}}(4) - \frac{1}{8}(25)^{-\frac{3}{2}}(4)^2$$

$$+ \frac{1}{16}(25)^{-\frac{5}{2}}(4)^3 - \cdots$$

$$= 5 + \left(\frac{1}{2}\right)\left(\frac{1}{5}\right)(4) - \left(\frac{1}{8}\right)\left(\frac{1}{125}\right)(16)$$

$$+ \left(\frac{1}{16}\right)\left(\frac{1}{3125}\right)(64) - \cdots$$

$$= 5 + 0.4 - 0.016 + 0.0013 - \cdots$$

$$\approx 5.385 \text{ (to three decimal places)}$$

Notice that the expansion is infinite, but we needed only the first four terms to evaluate the radical to three decimal places.

The key to this type of example, of course, is to find the sum of two numbers, one of which is a perfect power of the desired root.

EXAMPLES

$$\sqrt[3]{30} = \sqrt[3]{27 + 3} = (27 + 3)^{\frac{1}{3}}$$
$$\sqrt[5]{38} = \sqrt[5]{32 + 6} = (32 + 6)^{\frac{1}{5}}$$

By this method, any root of any number can be found to the desired degree of accuracy.

Exercise 8.3.1

Expand the following by the binomial theorem. Repeat the expansion using the alternate methods discussed in this section.

1. $(x + 2)^4$
2. $(1 + 2x)^6$
3. $(x - y)^6$
4. $(2x^2 - y)^5$
5. $\left(2a + \frac{b}{2}\right)^7$

Use the binomial theorem to find the indicated term or terms in each of the following:

6. $(x + y)^{30}$, first four terms
7. $(a - 3b)^{100}$, first four terms

8.3 THE BINOMIAL THEOREM

8. $\left(x - \dfrac{2}{x}\right)^{10}$, first four terms 9. $(2a^2 - b^2)^7$, fourth term

10. $(2x - \sqrt{y})^9$, seventh term

Find the exact value of each of the following using the binomial expansion:

11. 102^4 [*Hint:* $102^4 = (100 + 2)^4$]   12. 99^6

Use the binomial expansion to evaluate each of the following to three decimal places. Check your answer by using a calculator or logarithms.

13. $(1.02)^8$
14. $(0.97)^6$
15. $\sqrt{50}$
16. $\sqrt[3]{29}$
17. $\sqrt[4]{1.02}$
18. $(1.03)^{-6}$
19. $(1.99)^{-5}$

20. Show that $\dbinom{n}{r} = \dbinom{n}{r-1} \dfrac{n-(r-1)}{r}$.

8.4 AN INTRODUCTION TO PROBABILITY

Goals

Upon completion of this section you should be able to:
1. Find the probability of an independent event occurring.
2. Find the probability of two independent events occurring one after the other.
3. Given two events, determine the probability that either will occur.
4. Determine the odds of an event occurring.

* *

The idea of probability is expressed by most of us in our everyday language. "This will probably happen" and "the chances are that" are statements that are based to some degree on probability.

For our purposes here we need a precise definition of probability. This definition is based on events being *independent*. In other words, a particular outcome does not depend on a previous outcome. An example often used is the tossing of a coin. Whether the coin falls "heads" or "tails" does not depend on how it fell on the previous toss. The definition is also based on events being *equally likely*.

Definition

If an event can occur in any one of n equally likely and independent ways and if m of these ways are considered favorable, the *probability* of a favorable event is the ratio $\frac{m}{n}$.

Suppose we designate A as an event, then $P(A)$ is read as "the probability of A."

Our definition states

$$P(\text{favorable outcome}) = \frac{\text{number of favorable outcomes}}{\text{total number of outcomes}}$$

Suppose there are no favorable outcomes possible, then the ratio would be $\frac{0}{n}$ or 0. Also, if every outcome is favorable, then the ratio would be $\frac{n}{n}$ or 1. It is obvious that the numerator of the fraction cannot be larger than the denominator. This establishes that the probability of an event is never less than 0 nor greater than 1, that is

$$0 \leqslant P(A) \leqslant 1$$

EXAMPLE

Consider a drawing for a prize where there are ten cards with one name on each card. If Jay's name is on only one of the cards, then the chance of Jay's winning the prize is one in ten or

$$P(\text{Jay wins}) = \frac{1}{10}$$

If Jay's name is on three of the cards, then

$$P(\text{Jay wins}) = \frac{3}{10}$$

EXAMPLE

A die is rolled. What is the probability of obtaining a 5 on the top face?

A die (plural dice) is a regular cube with six faces marked with dots 1 through 6 as follows:

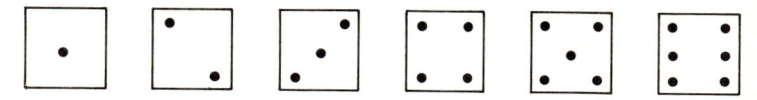

Only one face out of the six has a 5 on it. Thus

$$P(5) = \frac{1}{6}$$

8.4 AN INTRODUCTION TO PROBABILITY

EXAMPLE

A card is drawn at random from a bridge deck. What is the probability that the card is the ace of spades?

A bridge or poker deck is made up of four suits, two black suits (spades and clubs), and two red suits (hearts and diamonds). Each suit contains thirteen cards: nine of them numbered 2 through 10, plus three face cards (jack, queen, and king), and one ace.

There are thus 52 cards in a bridge deck and only one of them is the ace of spades. Hence

$$P(\text{ace of spades}) = \frac{1}{52}$$

EXAMPLE

If a card is drawn at random from a bridge deck, what is the probability of it being a club? There are 13 clubs in the deck of 52 cards, so

$$P(\text{club}) = \frac{13}{52} = \frac{1}{4}$$

EXAMPLE

A golf bag contains 3 red tees, 5 blue tees, and 7 white tees. What is the probability that a tee drawn at random will be blue? Since the bag contains a total of 15 tees and 5 are blue, the ratio of favorable outcomes (blue tee is drawn) to the total possible outcomes (15) is $\frac{5}{15}$ or $\frac{1}{3}$. Thus,

$$P(\text{blue}) = \frac{1}{3}$$

Notice that the probability of not drawing a blue tee in this example would be $\frac{10}{15}$ or $\frac{2}{3}$.

We use the notation $\overline{A}$ to designate "not A." Note that

$$P(A) + P(\overline{A}) = 1$$

or

$$P(A) = 1 - P(\overline{A})$$

This relationship is used in later formulas.

Exercise 8.4.1

1. If a coin is tossed, what is the probability that it will land heads up?
2. A die is rolled. What is the probability that the result will be

    a. a 3?   b. an even number?

    c. a number greater than 4?

3. If a card is drawn at random from a poker deck, what is the probability that it will be

    a. the king of clubs?   b. an ace?

    c. a heart?   d. a picture card?

4. Five hundred raffle tickets were sold for a prize, and you purchased 10 of them. What is the probability that you will win?

5. A candy jar contains 10 red jelly beans, 14 green jelly beans, 20 white jelly beans, and 18 black jelly beans. If a jelly bean is drawn at random from the jar, what is the probability that it will be

    a. red?   b. black?

    c. red or green?   d. not white?

    e. not green or black?

Two events are *independent* if the outcome of one does not affect the probability of the other. An example of independent events is the tossing of coins.

EXAMPLE

What is the probability of tossing two heads in succession with a single coin? If we examine this problem by listing the possibilities, we find there are four.

$$H H \quad H T \quad T H \quad T T$$

Out of the four possible outcomes there is one favorable event. Hence, by the definition of probability,

$$P(\text{two heads}) = \frac{1}{4}$$

Note also that the probability of obtaining a head on the first toss is $\frac{1}{2}$; the probability of obtaining a head on the second toss is $\frac{1}{2}$; and $\frac{1}{2} \times \frac{1}{2} = \frac{1}{4}$, which is the probability of two heads.

Theorem   If A and B are independent events, then $P(A \text{ and } B) = P(A) \times P(B)$.

EXAMPLE A die is rolled and a coin is tossed. What is the probability of obtaining a 6 on the die and a head on the coin?

$$P(6) = \frac{1}{6} \quad \text{and} \quad P(\text{head}) = \frac{1}{2}$$

$$P(6 \text{ and a head}) = \frac{1}{6} \times \frac{1}{2} = \frac{1}{12}$$

EXAMPLE A coin is tossed three times. What is the probability that it will land heads up all three times? On any given toss the probability of obtaining a head is $\frac{1}{2}$. Thus

$$P(\text{a head } and \text{ a head } and \text{ a head}) = \frac{1}{2} \times \frac{1}{2} \times \frac{1}{2} = \frac{1}{8}$$

EXAMPLE A die is rolled and three coins are tossed. What is the probability of obtaining a 4 on the die and 3 heads with the coins?

$$P(4) = \frac{1}{6}$$

$$P(3 \text{ heads}) = \frac{1}{2} \times \frac{1}{2} \times \frac{1}{2} = \frac{1}{8}$$

$$P(4 \text{ and 3 heads}) = \frac{1}{6} \times \frac{1}{8} = \frac{1}{48}$$

If two events are dependent (i.e., the outcome of one affects the probability of the other), we have a slightly different situation.

For example, suppose a card is drawn from a bridge deck, and then a second card is drawn without replacing the first card. If we wish to determine the probability that both cards are aces, we proceed as follows.

There are two events to consider.

1. The probability that the first card drawn is an ace.

$$P(\text{first card is an ace}) = \frac{4}{52} = \frac{1}{13}$$

2. Assuming that the first card is an ace (if it is not an ace, there is no need to consider the second card), we need to determine the probability that the second card is an ace. We have 51 cards remaining and 3 of them are aces. Therefore,

$$P(\text{second card is an ace}) = \frac{3}{51} = \frac{1}{17}$$

Thus the probability that both cards are aces is

$$P(\text{first is an ace } and \text{ second is an ace}) = \frac{1}{13} \times \frac{1}{17} = \frac{1}{221}$$

This illustrates the following theorem for dependent events.

Theorem **The probability of two events A and B occurring one after the other is the product of the probability of A and the probability of B, assuming A has occurred.**

$$P(A \text{ and } B) = P(A) \times P(B); \text{ assuming } A \text{ has occurred}$$

EXAMPLE A bag contains 5 red marbles, 7 green marbles, and 6 blue marbles. If 2 marbles are drawn, what is the probability of both being green?

$$P(\text{first green}) = \frac{7}{18}$$

$$P(\text{second green, assuming first is green}) = \frac{6}{17}$$

$$P(\text{both green}) = \frac{7}{18} \times \frac{6}{17} = \frac{7}{51}$$

EXAMPLE A die is rolled. What is the probability that the resulting number is odd and greater than 2?

$$P(\text{odd}) = \frac{3}{6} = \frac{1}{2}$$

$$P(\text{greater than 2, assuming it is odd}) = \frac{2}{3}$$

$$P(\text{odd and greater than 2}) = \frac{1}{2} \times \frac{2}{3} = \frac{1}{3}$$

Exercise 8.4.2

A calculator may be helpful in some of the following problems.

1. What is the probability of rolling two 6's with a pair of dice?
2. What is the probability of tossing five heads in succession with a single coin?

3. A die is rolled and a coin is tossed. What is the probability of obtaining an even number on the die and a tail on the coin?

4. A card is drawn at random from a bridge deck. It is replaced, the deck is shuffled, and a second card is drawn. What is the probability that

   a. both cards are aces?

   b. both cards are diamonds?

   c. both cards are picture cards?

   d. the first card is an ace and the second card is a club?

   e. both cards are greater than 7?

5. In a single roll of two dice, find the probability that the sum of the numbers obtained is even.

6. Two cards are drawn together from a bridge deck. What is the probability that

   a. both cards are jacks?

   b. both cards are hearts?

   c. both cards are picture cards?

   d. both cards are less than 9?

7. Two people are to be chosen at random from a group of four women and three men. What is the probability that they will both be women?

8. Six cards numbered 1 through 6 are placed in a box and three cards are drawn at random. What is the probability that they are all even numbers?

9. In a 12-horse race a person picks number 8 to win, number 4 to finish second, and number 10 to finish third. What is the probability that the horses will finish in exactly this order?

10. Five cards are dealt from a poker deck. What is the probability that they are all hearts?

11. A bridge hand consists of thirteen cards. What is the probability of being dealt all the spades?

Definition If two events cannot occur at the same time, they are said to be *mutually exclusive.*

An example of mutually exclusive events can demonstrate the formula for such probabilities.

268   THE BINOMIAL THEOREM AND AN INTRODUCTION TO PROBABILITY

EXAMPLE

A deck of cards is composed of 5 red cards, 3 blue cards, and 7 white cards. If a card is drawn at random, what is the probability that it will be red or blue?

This problem can be solved by noting that there are 15 possible outcomes. The 5 red and 3 blue cards (8 total) represent the favorable outcomes. Hence

$$P(\text{red or blue}) = \frac{8}{15}$$

Note also that
$$P(\text{red}) = \frac{5}{15}$$

and
$$P(\text{blue}) = \frac{3}{15}$$

and that
$$\frac{5}{15} + \frac{3}{15} = \frac{8}{15}$$

The formula for mutually exclusive events is

$$P(A \text{ or } B) = P(A) + P(B)$$

In logic, (A or B) is true in three cases: if A is true; if B is true; if both A and B are true. Since this definition of "or" allows both A and B to be true, it is important that we distinguish mutually exclusive events from those that can both happen at the same time.

If we wish to compute the probability of (A or B) when A and B are not mutually exclusive, we must be careful not to count an event twice.

EXAMPLE

If a single die is rolled, what is the probability of obtaining a number greater than 3 or an odd number?

$$P(\text{greater than 3}) = \frac{3}{6} = \frac{1}{2}$$

$$P(\text{odd number}) = \frac{3}{6} = \frac{1}{2}$$

If we add $\frac{1}{2} + \frac{1}{2}$, we get 1. This is obviously incorrect since a probability of 1 would exclude the probability of getting a 2, which is neither odd nor greater than 3. This discrepancy is caused by the fact

8.4 AN INTRODUCTION TO PROBABILITY

that 5 is both odd and greater than 3. We counted this probability twice.

$$P(5) = \frac{1}{6}$$

Our answer, therefore, should be

$$\frac{3}{6} + \frac{3}{6} - \frac{1}{6} = \frac{5}{6}$$

In general

$$P(A \text{ or } B) = P(A) + P(B) - P(A \text{ and } B)$$

Notice that this formula will also hold for mutually exclusive events, since $P(A \text{ and } B)$ is zero in that case.

EXAMPLE

A die is rolled and three coins are tossed. What is the probability of obtaining a 4 or obtaining 3 heads?

$$P(4) = \frac{1}{6} \quad \text{and} \quad P(3 \text{ heads}) = \frac{1}{8}$$

Also

$$P(4 \text{ and } 3 \text{ heads}) = \frac{1}{6} \times \frac{1}{8} = \frac{1}{48}$$

$$P(4 \text{ or } 3 \text{ heads}) = P(4) + P(3 \text{ heads}) - P(4 \text{ and } 3 \text{ heads})$$

$$= \frac{1}{6} + \frac{1}{8} - \frac{1}{48} = \frac{13}{48}$$

Exercise 8.4.3

1. A bag contains 5 white, 6 green, and 12 red marbles. If a marble is drawn at random, what is the probability that it will be red or green?
2. If a die is tossed, what is the probability that it will show a 3 or 5?
3. A card is drawn from a bridge deck. What is the probability that it is an ace or king?
4. Two dice are rolled. What is the probability that the sum of the numbers obtained is 3?
5. Two dice are rolled. What is the probability that the sum of the numbers is 7?
6. Two coins are tossed. What is the probability of obtaining at least one head?

7. One card is drawn from a bridge deck. What is the probability that it is an 8 or a club?

8. One card is drawn from a bridge deck. What is the probability that it is a heart or a picture card?

9. Two dice are rolled and two coins are tossed. What is the probability of obtaining a 12 or two heads?

10. A family has two children.

    a. What is the probability that one of the children is a girl? (Assume the chances of the child being a girl or a boy are equally likely.)

    b. If one of the children is a boy, what is the probability that the other is a girl?

    c. If the oldest child is a boy, what is the probability that the other is a girl?

A word closely associated with probability is *odds*. This word designates the relative chances of a favorable event occurring.

Definition The *odds* in favor of event A is the ratio of the probability of A to the probability of not A.

$$\text{Odds in favor of } A = \frac{P(A)}{P(\overline{A})}$$

EXAMPLE If a die is rolled, what are the odds a 2 will occur on the top face?

$$P(2) = \frac{1}{6} \quad \text{and} \quad P(\overline{2}) = \frac{5}{6}$$

Therefore, the odds in favor of a 2 $= \dfrac{\frac{1}{6}}{\frac{5}{6}} = \dfrac{1}{5}$.

The odds in favor of rolling a 2 with a single die are 1 to 5, often written as 1:5.

EXAMPLE A card is drawn from a bridge deck. What are the odds that the card will be a face card? There are 52 possible outcomes, 12 of which are favorable. So

$$P(\text{face card}) = \frac{12}{52} = \frac{3}{13}$$

8.4 AN INTRODUCTION TO PROBABILITY

and
$$P(\text{not a face card}) = 1 - \frac{3}{13} = \frac{10}{13}$$

Thus, the odds in favor of a face card $= \dfrac{\frac{3}{13}}{\frac{10}{13}} = \dfrac{3}{10}$.

Note also that the odds *against* drawing a face card would be 10:3.

EXAMPLE The odds in favor of an event A are 3:7. What is $P(A)$? From the formula we have

$$\frac{P(A)}{P(\overline{A})} = \frac{3}{7}$$

or
$$\frac{P(A)}{1 - P(A)} = \frac{3}{7}$$

Solving for $P(A)$ gives

$$P(A) = \frac{3}{10}$$

Exercise 8.4.4

A calculator may be helpful in some of the following problems.

1. If a die is rolled, what are the odds that a 5 will occur on the top face?
2. A die is rolled. What are the odds that an even number will occur on the top face?
3. A card is drawn from a bridge deck. What are the odds that the card will be an ace?
4. What are the odds of obtaining three heads on a toss of three coins?
5. Two dice are rolled. What are the odds that the sum of the faces will be

    a. 2? b. 7?

6. A box contains 6 blue cards, 10 red cards, 7 green cards, and 9 white cards. If a card is drawn at random, what are the odds that it will be

    a. red? b. blue or red? c. not blue or green?

7. Two cards are drawn from a bridge deck. What are the odds that they both will be aces?

8. The odds against a horse winning a race are 9:5. What is the probability that the horse will win the race?

9. Five cards are dealt from a poker deck. What are the odds that they will all be clubs?

10. On a roulette wheel there are 38 positions in which the ball can stop (numbers 1 to 36, plus 0 and 00). There are 18 red numbers and 18 black numbers (the 0 and 00 positions are green). If the wheel is spun, what are the odds that the ball will stop at

   a. 00?
   b. a red number?
   c. a green number?
   d. a red or black number?

CHAPTER REVIEW

A calculator may be helpful in some of the following problems.

1. Evaluate 8!.

2. Evaluate $\dfrac{18!}{15!}$.

3. Evaluate $_{10}P_3$.

4. Evaluate $\binom{12}{4}$.

5. In how many ways can the letters of the word "trapezoid" be arranged?

6. The three offices of president, vice president, and secretary are to be elected from a group of 35 people. If each person is eligible for election to any office, how many possible arrangements are there?

7. How many different seven-digit telephone numbers can be formed from the digits 0 through 9 if no digit is used more than once?

8. How many different seven-digit telephone numbers can be formed from the digits 0 through 9 if digits may be repeated?

9. How many triangles are determined by 12 points, no 3 of which are collinear?

10. In how many ways can a set of 13 cards be selected from a deck of 52 cards?

11. A person is to select one or more books from a set of seven books. How many possible combinations are there?

12. A committee of three is to be selected from a group of ten teachers and twelve parents. If the committee must have at least one teacher as a member, how many possible committees could be formed?

13. Expand $(2x + 3y)^6$ by using the binomial theorem.
14. Expand $(y^3 - 3a)^4$ by using the binomial theorem.
15. Find the fifth term of the expansion of $(y^2 - 2)^{10}$.
16. Find the middle term of the expansion of $\left(x - \dfrac{2}{x}\right)^8$.
17. Evaluate $(0.8)^5$ using the binomial expansion.
18. Use the binomial expansion to evaluate $\sqrt{17}$ to three decimal places.
19. Evaluate $\sqrt[4]{19}$ to three decimal places using the binomial expansion.
20. A single die is rolled. What is the probability that the top face will be greater than 4?
21. A die is rolled and a card is drawn from a bridge deck. What is the probability that the top face of the die will be a 6 and the card will be an ace?
22. If three coins are tossed, what is the probability that at least one will land tails? (*Hint:* Find the probability that all three will not be heads.)
23. A box contains 11 red, 4 green, 6 blue, and 10 yellow cards. If a card is drawn at random, what is the probability that it is either green or yellow?
24. A card is drawn from a bridge deck. What is the probability that it is a club or greater than 10?
25. Five coins are tossed. What are the odds of obtaining all heads?
26. Five dice are rolled. Find the odds that all faces will be the same value.
27. Five cards are dealt from a poker deck. What are the odds against a "royal flush" (i.e., the 10, jack, queen, king, and ace of the same suit) being dealt?

PRACTICE TEST

1. Evaluate $\dfrac{9!\,5!}{10!}$.
2. Evaluate $_{20}P_4$.
3. Evaluate $\dbinom{18}{3}$.
4. The four positions of president, vice president, secretary, and treasurer are to be selected from a class of 30. In how many ways can these offices be filled?
5. How many different committees of four persons each can be chosen from a group of ten persons?

6. Find the tenth term of $\left(x + \dfrac{1}{2y}\right)^{12}$.

7. Use the binomial expansion to evaluate $\sqrt{15}$ to two decimal places.

8. Eight cards numbered 1 to 8 are placed in a box and three cards are drawn at random. What is the probability that they are all even numbers?

9. A card is drawn from a bridge deck. What is the probability that the card is a club or less than 7? (Assume that an ace is a high card.)

10. A die is rolled and a coin is tossed. What are the odds of obtaining a head and an even number?

Common logarithms

N	0	1	2	3	4	5	6	7	8	9
10	0000	0043	0086	0128	0170	0212	0253	0294	0334	0374
11	0414	0453	0492	0531	0569	0607	0645	0682	0719	0755
12	0792	0828	0864	0899	0934	0969	1004	1038	1072	1106
13	1139	1173	1206	1239	1271	1303	1335	1367	1399	1430
14	1461	1492	1523	1553	1584	1614	1644	1673	1703	1732
15	1761	1790	1818	1847	1875	1903	1931	1959	1987	2014
16	2041	2068	2095	2122	2148	2175	2201	2227	2253	2279
17	2304	2330	2355	2380	2405	2430	2455	2480	2504	2529
18	2553	2577	2601	2625	2648	2672	2695	2718	2742	2765
19	2788	2810	2833	2856	2878	2900	2923	2945	2967	2989
20	3010	3032	3054	3075	3096	3118	3139	3160	3181	3201
21	3222	3243	3263	3284	3304	3324	3345	3365	3385	3404
22	3424	3444	3464	3483	3502	3522	3541	3560	3579	3598
23	3617	3636	3655	3674	3692	3711	3729	3747	3766	3784
24	3802	3820	3838	3856	3874	3892	3909	3927	3945	3962
25	3979	3997	4014	4031	4048	4065	4082	4099	4116	4133
26	4150	4166	4183	4200	4216	4232	4249	4265	4281	4298
27	4314	4330	4346	4362	4378	4393	4409	4425	4440	4456
28	4472	4487	4505	4518	4533	4548	4564	4579	4594	4609
29	4624	4639	4654	4669	4683	4698	4713	4728	4742	4757
30	4771	4786	4800	4814	4829	4843	4857	4871	4886	4900
31	4914	4928	4942	4955	4969	4983	4997	5011	5024	5038
32	5051	5065	5079	5092	5105	5119	5132	5145	5159	5172
33	5185	5198	5211	5224	5237	5250	5263	5276	5289	5302
34	5315	5328	5340	5353	5366	5378	5391	5403	5416	5428
35	5441	5453	5465	5478	5490	5502	5514	5527	5539	5551
36	5563	5575	5587	5599	5611	5623	5635	5647	5658	5670
37	5682	5694	5705	5717	5729	5740	5752	5763	5775	5786
38	5798	5809	5821	5832	5843	5855	5866	5877	5888	5899
39	5911	5922	5933	5944	5955	5966	5977	5988	5999	6010
40	6021	6031	6042	6053	6064	6075	6085	6096	6107	6117
41	6128	6138	6149	6160	6170	6180	6191	6201	6212	6222
42	6232	6243	6253	6263	6274	6284	6294	6304	6314	6325
43	6335	6345	6355	6365	6375	6385	6395	6405	6415	6425
44	6435	6444	6454	6464	6474	6484	6493	6503	6513	6522
45	6532	6542	6551	6561	6571	6580	6590	6599	6609	6618
46	6628	6637	6646	6656	6665	6675	6684	6693	6702	6712
47	6721	6730	6739	6749	6758	6767	6776	6785	6794	6803
48	6812	6821	6830	6839	6848	6857	6866	6875	6884	6893
49	6902	6911	6920	6928	6937	6946	6955	6964	6972	6981
50	6990	6998	7007	7016	7024	7033	7042	7050	7059	7067
51	7076	7084	7093	7101	7110	7118	7126	7135	7143	7152
52	7160	7168	7177	7185	7193	7202	7210	7218	7226	7235
53	7243	7251	7259	7267	7275	7284	7292	7300	7308	7316
54	7324	7332	7340	7348	7356	7364	7372	7380	7388	7396

Common logarithms

N	0	1	2	3	4	5	6	7	8	9
55	7404	7412	7419	7427	7435	7443	7451	7459	7466	7474
56	7482	7490	7497	7505	7513	7520	7528	7536	7543	7551
57	7559	7566	7574	7582	7589	7597	7604	7612	7619	7627
58	7634	7642	7649	7657	7664	7672	7679	7686	7694	7701
59	7709	7715	7723	7731	7738	7745	7752	7760	7767	7774
60	7782	7789	7796	7803	7810	7818	7825	7832	7839	7846
61	7853	7860	7868	7875	7882	7889	7896	7903	7910	7917
62	7924	7931	7938	7945	7952	7959	7966	7973	7980	7987
63	7993	8000	8007	8014	8021	8028	8035	8041	8048	8055
64	8062	8069	8075	8082	8089	8096	8102	8109	8116	8122
65	8129	8136	8142	8149	8156	8162	8169	8176	8182	8189
66	8195	8202	8209	8215	8222	8228	8235	8241	8248	8254
67	8261	8267	8274	8280	8287	8293	8299	8306	8312	8319
68	8325	8331	8338	8344	8351	8357	8363	8370	8376	8382
69	8388	8395	8401	8407	8414	8420	8426	8432	8439	8445
70	8451	8457	8563	8470	8476	8482	8488	8494	8500	8506
71	8513	8519	8525	8531	8537	8543	8549	8555	8561	8567
72	8573	8579	8585	8591	8597	8603	8609	8615	8621	8627
73	8633	8639	8645	8651	8657	8663	8669	8675	8681	8686
74	8692	8698	8704	8710	8716	8722	8727	8733	8739	8745
75	8751	8756	8762	8768	8774	8779	8785	8791	8797	8802
76	8808	8814	8820	8825	8831	8837	8842	8848	8854	8859
77	8865	8871	8876	8882	8887	8893	8899	8904	8910	8915
78	8921	8927	8932	8938	8943	8949	8954	8960	8965	8971
79	8976	8982	8987	8993	8998	9004	9009	9015	9020	9025
80	9031	9036	9042	9047	9053	9058	9063	9069	9074	9079
81	9085	9090	9096	9101	9106	9112	9117	9122	9128	9133
82	9138	9143	9149	9154	9159	9165	9170	9175	9180	9186
83	9191	9196	9201	9206	9212	9217	9222	9227	9232	9238
84	9243	9248	9253	9258	9263	9269	9274	9279	9284	9289
85	9294	9299	9304	9309	9315	9320	9325	9330	9335	9340
86	9345	9350	9355	9360	9365	9370	9375	9380	9385	9390
87	9395	9400	9405	9410	9415	9420	9425	9430	9435	9440
88	9445	9450	9455	9460	9465	9469	9474	9479	9484	9489
89	9494	9499	9504	9509	9513	9518	9523	9528	9533	9538
90	9542	9547	9552	9557	9562	9566	9571	9576	9581	9586
91	9590	9595	9600	9605	9609	9614	9619	9624	9628	9633
92	9638	9643	9647	9652	9657	9661	9666	9671	9675	9680
93	9685	9689	9694	9699	9703	9708	9713	9717	9722	9727
94	9731	9736	9741	9745	9750	9754	9759	9763	9768	9773
95	9777	9782	9786	9791	9795	9800	9805	9809	9814	9818
96	9823	9827	9832	9836	9841	9845	9850	9854	9859	9863
97	9868	9872	9877	9881	9886	9890	9894	9899	9903	9908
98	9912	9917	9921	9926	9930	9934	9939	9943	9948	9952
99	9956	9961	9965	9969	9974	9978	9983	9987	9991	9996

Answers to Odd-numbered Exercises

CHAPTER 1

1.1.1 (page 5)
1. $4 < 6$ 3. $-3 < 0$ 5. $-1 > -5$ 7. $\frac{3}{2} < 2$ 9. $-0.5 > -0.6$

1.1.2 (page 6)
1. a. 6   b. 1   c. 0   d. 3   e. 3   f. 13   g. $\frac{1}{40}$ h. $\frac{1}{63}$
3. (a) a, b, c, d, e, g   (b) a, b, c, e, f, g, h   5. $a > 0, a < 0, a = 0$

1.2.1 (page 9)
1. $16x^8$ 3. $\frac{8}{x^6 y^3}$ 5. $\frac{y^2}{x^2}$ 7. $8x^6 y^6$ 9. 1   11. $-\frac{1}{2x}$
13. $2x^8 y^4$ 15. $288 x^{22} y^{16}$ 17. $-\frac{2}{x}$ 19. $-\frac{1}{10xy^4}$ 21. $\frac{1}{x^3 y^5}$
23. 9   25. $\frac{1}{a+b}$ 27. $\frac{y^4}{x^2 z}$ 29. $\frac{x^8}{y^{16}}$ 31. $\frac{1}{x^{15}}$ 33. $\frac{1}{y^7}$
35. $-\frac{9}{8x^{11}}$ 37. $\frac{x^5}{y^{25}}$ 39. $\frac{y^3}{x^3}$

1.2.2 (page 15)
1. 12   3. -2 5. -1 7. $5\sqrt{5}$ 9. $x^2 y^3$ 11. $\sqrt[6]{x^5 y^3}$
13. $4xy^2 \sqrt[3]{x}$ 15. $y^2 \sqrt[3]{5xy}$ 17. $7\sqrt{5} - 3\sqrt{3}$ 19. $3\sqrt{2} - 3$
21. $\sqrt{2} - \sqrt[3]{2}$ 23. $6\sqrt{14}$ 25. $30\sqrt{2} + 15\sqrt{5}$ 27. $\sqrt{21} + \sqrt{10}$
29. -167 31. $\frac{3\sqrt{5}}{10}$ 33. $\frac{\sqrt[3]{2xy^2}}{x}$ 35. $\frac{8\sqrt{x-2}}{x-2}$ 37. $3 + \sqrt{5}$
39. $\sqrt{a} + 1$

279

1.3.1 (page 17)
1. 20   3. −25   5. $\dfrac{16}{3}$ 7. $\dfrac{3}{2}$ 9. 65

1.3.2 (page 19)
1. yes, yes, yes, yes   3. no, no, no, yes   5. yes, yes, yes, yes
7. no, no, yes, no

1.3.3 (page 20)
1. $2x^2y + 3xy^2$ 3. $2x^2 - xy + 6y^2$ 5. $ax + bx - a - b$
7. $x^3 + x^2 - 3x + 9$ 9. $x^3 + 27$ 11. $a^2 - b^2$ 13. $a^3 + b^3$

1.4.1 (page 23)
1. $5a(a - 2)$ 3. $(3x - 5)^2$ 5. $5(x + 3)(x - 3)$ 7. $(x + 5)(x - 3)$
9. $(2x - 3)(x + 1)$ 11. $(2x + 1)(3x - 4)$ 13. $2x(x + 2)(3x - 2)$
15. prime   17. $(x - 5)(a - 4)$ 19. $(2x + 3)(a + 1)(a - 1)$
21. $(x - y^2)(x^2 + xy^2 + y^4)$ 23. $(2x + 5)(x + 9)$
25. $(x + y + 3)(x - y - 1)$ 27. $(3a + 3x + 4y)(3a - 3x - 4y)$
29. $(a + 3)(a - 3)(3b - 2)$

1.5.1 (page 26)
1. $\dfrac{x + 3}{x + 8}$ 3. $\dfrac{3}{5(x - 1)}$ 5. $x + 2$ 7. $\dfrac{1}{(x + 1)(x - 1)}$ 9. $\dfrac{x + 2}{x + 4}$
11. $\dfrac{x^2 + 4}{x(x + 3)}$ 13. $\dfrac{2x - 3}{(x - 1)(x - 2)}$ 15. $\dfrac{x^2 + 6x - 8}{(x + 2)(x - 2)(x + 6)}$
17. $\dfrac{1}{a}$ 19. $-x^2 + x + 6$

Chapter 1 Review (page 27)
1. 4   3. commutative property of addition   5. multiplicative inverse
7. $4x^8y^9z^8$ 9. $\dfrac{x^8}{y^{12}}$ 11. not a real number   13. $y\sqrt[4]{4x^3y}$
15. $6\sqrt{3} - 8\sqrt{5} - 24$ 17. $\dfrac{\sqrt{3}}{6}$ 19. $-\dfrac{5}{3}$ 21. $-x - 2$
23. $5x^2y(3x^2 - 4y + 1)$ 25. $(x + 15)(x + 1)$ 27. $(x + 28)(x - 5)$
29. $(3x - 2)^2$ 31. $4x(x + 5)(x - 3)$ 33. $(x - 3)(a^2 + 1)(a - 1)(a + 1)$
35. $\dfrac{3a + 2}{2a - 3}$ 37. $-\dfrac{1}{2x + 3}$ 39. $\dfrac{2a + b}{a^2 + 2a + ab}$

Chapter 1 Practice Test (page 28)

1. $\dfrac{11}{56}$

3. a. $\dfrac{1}{729}$ b. -1 c. $\sqrt[5]{4x^3y^4}$
   d. $\sqrt{2}$ e. $12\sqrt{2} - 12 + 12\sqrt{6}$ f. $\dfrac{\sqrt{5}-1}{2}$

5. a. $5a^2b + 5a^2b^2$ b. $3y$ c. $x^3 - 125$

7. a. $\dfrac{x+7}{(x+9)(x-4)}$ b. $\dfrac{x+5}{x+3}$ c. $\dfrac{xy-x^2}{y}$

CHAPTER 2

2.1.1 (page 34)

1. 19 3. $\dfrac{14}{5}$ 5. $\dfrac{45}{4}$ 7. $\dfrac{25}{6}$ 9. $-\dfrac{1}{5}$

2.1.2 (page 35)

1. 10 3. $-\dfrac{9}{8}$ 5. no solution 7. 4 9. -2 11. no solution
13. 3 15. 0 17. -2 19. 5

2.2.1 (page 39)

1. $\{-3, 0\}$ 3. $\{-3, 1\}$ 5. $\{3\}$ 7. $\left\{\dfrac{1}{5}, \dfrac{4}{3}\right\}$ 9. $\left\{-2y, \dfrac{y}{3}\right\}$ 11. $\left\{\dfrac{3}{2}\right\}$

2.2.2 (page 41)

1. $\{-5, 3\}$ 3. $\left\{0, -\dfrac{3}{2}\right\}$ 5. $\left\{\dfrac{-7 \pm \sqrt{29}}{10}\right\}$ 7. $\{-1 \pm 2\sqrt{2}\}$
9. $\left\{\dfrac{-4 \pm \sqrt{31}}{3}\right\}$ 11. 1, unequal, rational 13. 5, unequal, irrational
15. -11, no real roots 17. 625, unequal, rational 19. 49, unequal, rational

2.3.1 (page 46)

1. $7 + 7i$ 3. $-3 - i$ 5. $5 + 16i$ 7. $x^2 + y^2$ 9. $11 - 60i$
11. $\dfrac{4 - 3i}{5}$

2.3.2 (page 47)

1. $\{2 \pm 2i\}$ 3. $\left\{\dfrac{1 \pm i}{4}\right\}$ 5. $\left\{\dfrac{1 \pm i\sqrt{3}}{2}\right\}$ 7. $\left\{1, \dfrac{-1 \pm i\sqrt{3}}{2}\right\}$

9. $\left\{3, \dfrac{-3 \pm 3i\sqrt{3}}{2}\right\}$ 11. $\{\pm 2, \pm 2i\}$

13. No. The field of complex numbers is not ordered.

2.4.1 (page 51)
1. $\{\pm 1, \pm 2\}$ 3. $\{\pm\sqrt{2}, \pm\sqrt{5}\}$ 5. $\{\pm i\sqrt{3}, \pm i\sqrt{5}\}$ 7. $\{16, 81\}$
9. $\{9\}$ 11. $\{0, 4\}$ 13. $\{7\}$ 15. $\{-3\}$
17. $\left\{3, -\dfrac{7}{5}\right\}$ 19. $\left\{-5, 2, \dfrac{-3 \pm \sqrt{33}}{2}\right\}$ 21. $\{49\}$ 23. no solution
25. $\{2\}$ 27. no solution   29. $\left\{-\dfrac{1}{2}, 5\right\}$

2.5.1 (page 53)
1. $x \in [3, 8]$ 3. $x \in (-1, 6)$ 5. $x \in (-4, -2]$

7.

9.

2.5.2 (page 55)
1. $\left(-\infty, -\dfrac{1}{2}\right)$ 3. $(24, +\infty)$ 5. $(-\infty, 0)$ 7. $(-3, 3)$
9. $\left(-\infty, -\dfrac{4}{7}\right] \cup [2, +\infty)$

11.

13.

15.

17.

19. [number line with marks at $-\frac{1}{3}$, 0, and 3]

2.6.1 (page 58)
1. $(2,3)$ 3. $(-\infty, -3) \cup (4, +\infty)$ 5. $\left[-\frac{3}{2}, 6\right]$ 7. $(-\infty, -1) \cup (2,4)$
9. $(-\infty, 2] \cup (3,5]$ 11. no solution

2.7.1 (page 62)
1. 28, 30   3. 3 meters   5. $6,400 at 6%; $8,600 at 8%
7. 4 kilometers per hour   9. 1 hour, 60 kilometers   11. 21 liters
13. $5,000 at each rate

Chapter 2 Review (page 63)
1. $\{-3\}$ 3. $\{1\}$ 5. no solution   7. $\{-7, -3\}$ 9. $\{-2\}$
11. 81, real, rational, unequal   13. -7, no real roots   15. $9 + 2i$
17. $14 + 13i$ 19. $-2i$ 21. $\{1 \pm 2i\}$ 23. $\left\{5, \dfrac{-5 \pm 5i\sqrt{3}}{2}\right\}$
25. $\{16, 625\}$ 27. $\{-64, 1\}$ 29. $\{-6\}$ 31. $\left\{-5, 2, \dfrac{-3 \pm \sqrt{17}}{2}\right\}$
33. $(-\infty, 3]$ 35. $\left(-\dfrac{5}{3}, 3\right)$ 37. $(-7, 3)$ 39. $(-\infty, 5) \cup (5, +\infty)$
41. $57°, 37°, 86°$

Chapter 2 Practice Test (page 64)
1. a. 17, unequal, irrational   b. 49, unequal, rational   c. -44, no real roots
3. a. $\{-7\}$ b. $\left\{\dfrac{5}{2}\right\}$ c. $\left\{\dfrac{1}{3}\right\}$ d. $\left\{-2, \dfrac{1}{2}\right\}$ e. $\left\{-\dfrac{3}{2}\right\}$
f. $\left\{\dfrac{3 \pm \sqrt{5}}{2}\right\}$ g. $\{1 \pm i\}$ h. $\left\{1, \dfrac{-1 \pm i\sqrt{3}}{2}\right\}$ i. $\{\pm 1, \pm 3\}$
j. $\{5\}$ k. $\left\{\dfrac{-5 \pm \sqrt{5}}{2}, \dfrac{-5 \pm i\sqrt{11}}{2}\right\}$ l. $\left(-\infty, -\dfrac{31}{4}\right)$ m. $\left[-\dfrac{3}{5}, 1\right]$
n. $(-3, 1) \cup (2, +\infty)$

CHAPTER 3

3.1.1 (page 71)
1. a. $(-\infty, +\infty)$ b. $[-1, +\infty)$ c. yes
3. a. $[-3, +\infty)$ b. $(-\infty, +\infty)$ c. no
5. a. $(-\infty, +\infty)$ b. $(-\infty, +\infty)$ c. yes
7. a. $[-3, 3]$ b. $[0, 3]$ c. yes

ANSWERS TO ODD-NUMBERED EXERCISES

9. a. $[0, +\infty)$ b. $(-\infty, +\infty)$ c. no
11. a. $(-\infty, 3) \cup (3, +\infty)$ b. $(-\infty, 1) \cup (1, +\infty)$ c. yes
13. a. $(-\infty, 0) \cup (0, +\infty)$ b. $(-\infty, -1) \cup (1, +\infty)$ c. yes
15. a. $(-\infty, 0) \cup (0, +\infty)$ b. $\{-1, 1\}$ c. yes

3.2.1 (page 78)
1. a. 8 b. $8\sqrt{2}$ c. $\sqrt{13}$ d. $\sqrt{37}$
3. $\overline{AB} + \overline{BC} = 4\sqrt{2} + 3\sqrt{2} = 7\sqrt{2} = \overline{AC}$ 5. $(-5, 16)$

3.3.1 (page 82)
1. $(x + 2)^2 + (y - 3)^2 = 36$

3. a.

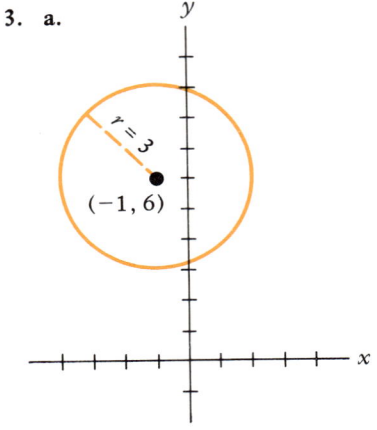

b.

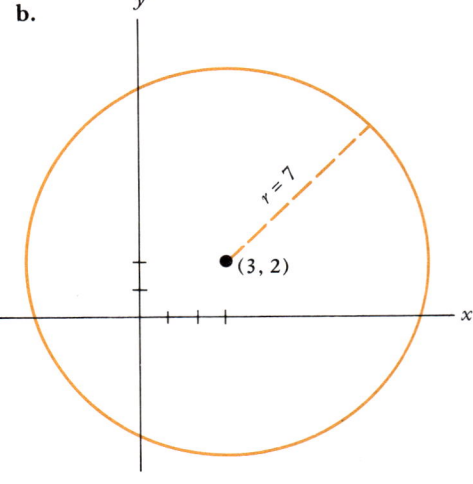

c.

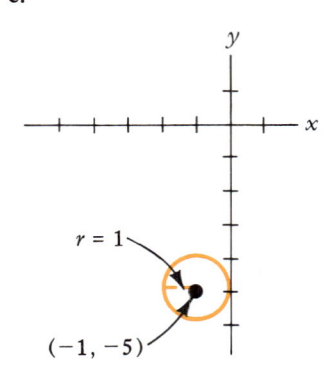

d.

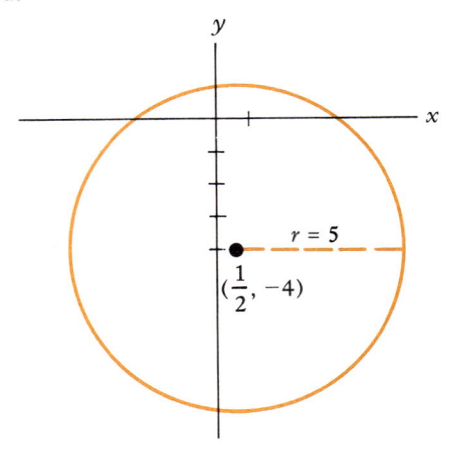

284 ANSWERS TO ODD-NUMBERED EXERCISES

5. $(x+1)^2 + (y-2)^2 = 137$

3.3.2 (page 86)

1. **a.**

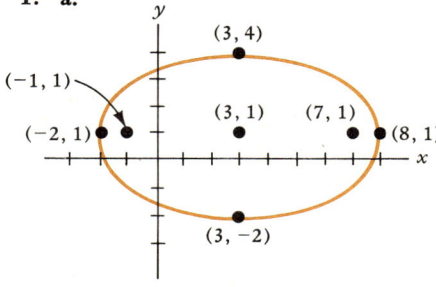

Points shown: $(3, 4)$, $(-1, 1)$, $(-2, 1)$, $(3, 1)$, $(7, 1)$, $(8, 1)$, $(3, -2)$

b.

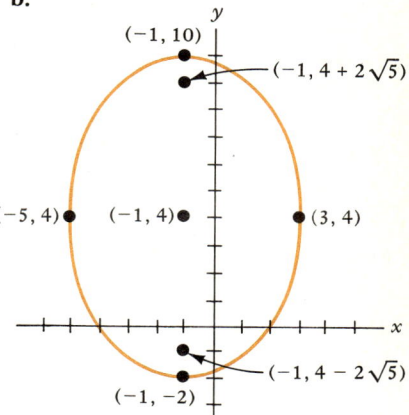

Points shown: $(-1, 10)$, $(-1, 4 + 2\sqrt{5})$, $(-5, 4)$, $(-1, 4)$, $(3, 4)$, $(-1, 4 - 2\sqrt{5})$, $(-1, -2)$

c.

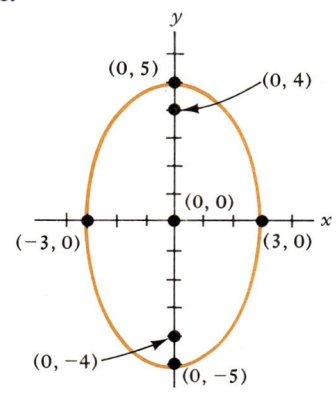

Points shown: $(0, 5)$, $(0, 4)$, $(0, 0)$, $(-3, 0)$, $(3, 0)$, $(0, -4)$, $(0, -5)$

d.

Points shown: $(2, -3)$, $(2 - 2\sqrt{3}, -5)$, $(2 + 2\sqrt{3}, -5)$, $(-2, -5)$, $(2, -5)$, $(6, -5)$, $(2, -7)$

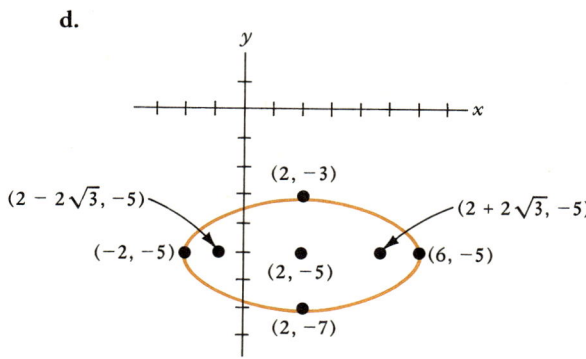

3. $\dfrac{(x+1)^2}{25} + \dfrac{(y-3)^2}{16} = 1$

3.3.3 (page 89)

1. a.

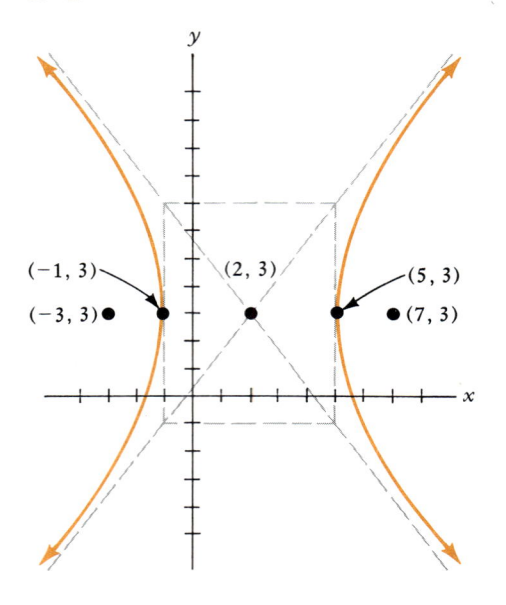

b.

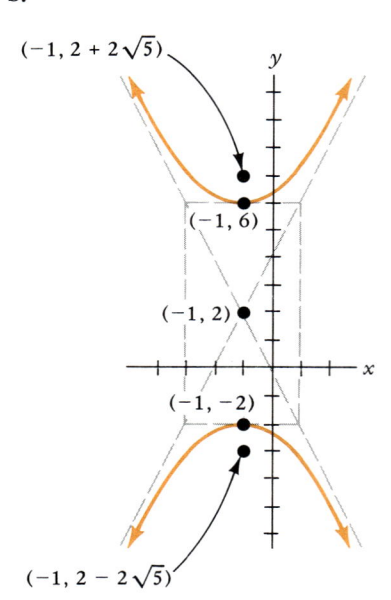

c.

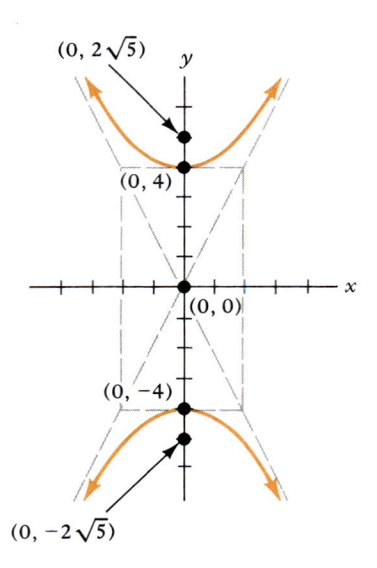

d.

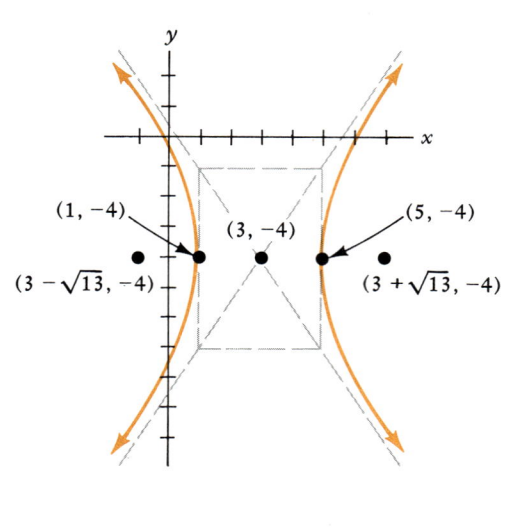

3. $\dfrac{(x-1)^2}{9} - \dfrac{(y-2)^2}{7} = 1$

3.3.4 (page 92)

1. a.

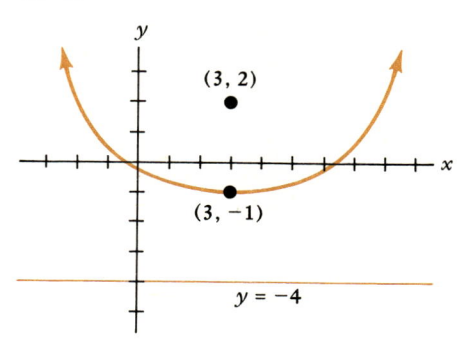

[Graph showing upward-opening parabola with vertex (3, −1), focus (3, 2), and directrix $y = -4$]

b.

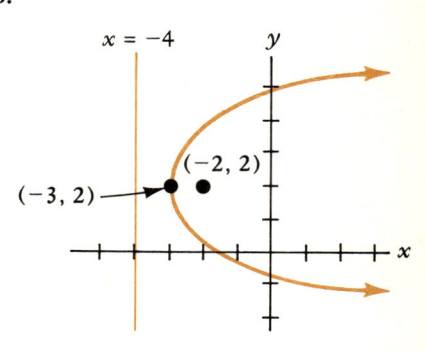

[Graph showing rightward-opening parabola with vertex (−3, 2), focus (−2, 2), and directrix $x = -4$]

c.

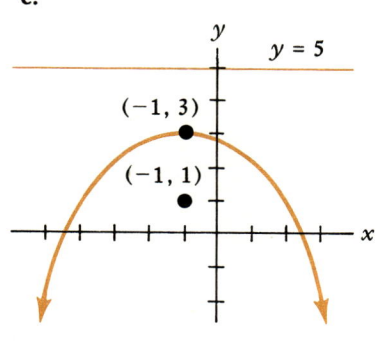

[Graph showing downward-opening parabola with vertex (−1, 3), focus (−1, 1), and directrix $y = 5$]

d.

[Graph showing leftward-opening parabola with vertex (2, 1), focus (0, 1), and directrix $x = 4$]

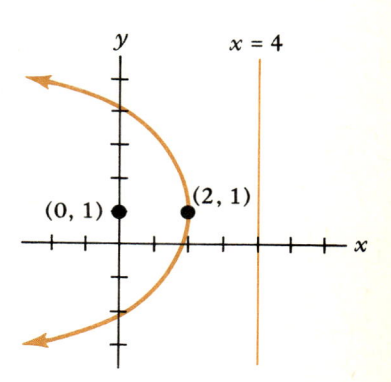

3. $(x-3)^2 = 12(y-1)$

3.4.1 (page 97)

1. a. $8x + 5y = 21$ b. $x + 2y = -2$ c. $5x + 12y = 20$
   d. $4x + 11y = -43$

3. a. $y = -2x + 5$, $m = -2$, $b = 5$ b. $y = -\dfrac{3}{4}x + \dfrac{1}{4}$, $m = -\dfrac{3}{4}$, $b = \dfrac{1}{4}$

   c. $y = \dfrac{2}{5}x - \dfrac{9}{5}$, $m = \dfrac{2}{5}$, $b = -\dfrac{9}{5}$ d. $y = 2x + \dfrac{4}{3}$, $m = 2$, $b = \dfrac{4}{3}$

e. $y = \dfrac{5}{2}x$, $m = \dfrac{5}{2}$, $b = 0$ f. $y = \dfrac{1}{8}x - \dfrac{1}{8}$, $m = \dfrac{1}{8}$, $b = -\dfrac{1}{8}$

5. a. $2x - y = 5$ b. $3x + y = 6$ c. $5x - 2y = 27$ d. $2x - y = 16$
   e. $y = 1$ f. $x = 4$

3.5.1 (page 99)

1. a. $(f + g)(x) = x^2 + 3x + 2$, $(-\infty, +\infty)$
   b. $(f - g)(x) = x^2 - 3x - 2$, $(-\infty, +\infty)$
   c. $(f \cdot g)(x) = 3x^3 + 2x^2$, $(-\infty, +\infty)$
   d. $\left(\dfrac{f}{g}\right)(x) = \dfrac{x^2}{3x + 2}$, $\left(-\infty, -\dfrac{2}{3}\right) \cup \left(-\dfrac{2}{3}, +\infty\right)$
   e. $\left(\dfrac{g}{f}\right)(x) = \dfrac{3x + 2}{x^2}$, $(-\infty, 0) \cup (0, +\infty)$

3. a. $(f + g)(x) = \sqrt{x} + x^2$, $[0, +\infty)$ b. $(f - g)(x) = \sqrt{x} - x^2$, $[0, +\infty)$
   c. $(f \cdot g)(x) = x^2\sqrt{x}$, $[0, +\infty)$ d. $\left(\dfrac{f}{g}\right)(x) = \dfrac{\sqrt{x}}{x^2}$, $(0, +\infty)$
   e. $\left(\dfrac{g}{f}\right)(x) = \dfrac{x^2}{\sqrt{x}}$, $(0, +\infty)$

5. a. $(f + g)(x) = \dfrac{x^2 + 2x + 4}{x(x + 4)}$, $(-\infty, -4) \cup (-4, 0) \cup (0, +\infty)$
   b. $(f - g)(x) = \dfrac{4 - x^2}{x(x + 4)}$, $(-\infty, -4) \cup (-4, 0) \cup (0, +\infty)$
   c. $(f \cdot g)(x) = \dfrac{x + 1}{x(x + 4)}$, $(-\infty, -4) \cup (-4, 0) \cup (0, +\infty)$
   d. $\left(\dfrac{f}{g}\right)(x) = \dfrac{x + 4}{x(x + 1)}$, $(-\infty, -4) \cup (-4, -1) \cup (-1, 0) \cup (0, +\infty)$
   e. $\left(\dfrac{g}{f}\right)(x) = \dfrac{x(x + 1)}{x + 4}$, $(-\infty, -4) \cup (-4, 0) \cup (0, +\infty)$

3.5.2 (page 100)

1. $(f \circ g)(x) = 10x - 14$, $(-\infty, +\infty)$ $(g \circ f)(x) = 10x - 1$, $(-\infty, +\infty)$
3. $(f \circ g)(x) = 8x^2 + 6x + 2$, $(-\infty, +\infty)$ $(g \circ f)(x) = 4x^2 - 2x + 3$, $(-\infty, +\infty)$
5. $(f \circ g)(x) = \sqrt{3x^2 + 8}$, $(-\infty, +\infty)$ $(g \circ f)(x) = 3x + 2$, $\left[\dfrac{1}{3}, +\infty\right)$
7. $(f \circ g)(x) = 8$, $(-\infty, +\infty)$ $(g \circ f)(x) = 2$, $(-\infty, +\infty)$
9. $(f \circ g)(x) = x$, $(-\infty, +\infty)$ $(g \circ f)(x) = x$, $(-\infty, +\infty)$

3.6.1 (page 102)

1. a. $(3, 1), (5, 2), (3, -2), (-1, 0)$ b. yes c. no
3. a. $y = \dfrac{x + 2}{5}$ b. yes c. yes

5. a. $y = \pm\sqrt{\dfrac{x+4}{3}}$ b. yes   c. no

7. a. $y = 2 \pm \sqrt{x-1}$ b. yes   c. no

3.6.2 (page 104)

1. a. $f^{-1}(x) = \dfrac{x+7}{2}$ b. no restrictions necessary

3. a. $f^{-1}(x) = \dfrac{1}{x}, x \neq 0$ b. no further restrictions necessary

5. a. $y = \pm\sqrt{x+3}$ b. $f(x) = x^2 - 3, x \geq 0; f^{-1}(x) = \sqrt{x+3}, x \geq -3$

7. a. $f^{-1}(x) = \sqrt[3]{x}$ b. no restrictions necessary

9. Yes. $f(x) = x^2, x \leq 0$ and $f^{-1}(x) = -\sqrt{x}, x > 0$ are inverse functions. $|x| = -x$ for $x \leq 0$.

Chapter 3 Review (page 105)

1. $D: (-\infty, +\infty), R: [0, +\infty)$, yes   3. $D: [1, +\infty), R: (-\infty, +\infty)$, no

5. $D: [-2, 2], R: [0, 2]$, yes   7. $D: (-\infty, -1) \cup (1, +\infty), R: (0, +\infty)$, yes

9. $2\sqrt{74}$ 11. $(x-3)^2 + (y+1)^2 = 25$, center: $(3, -1)$, radius: 5

13.

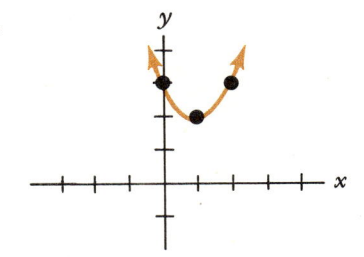

domain: $(-\infty, +\infty)$

range: $[2, +\infty)$

function

15.

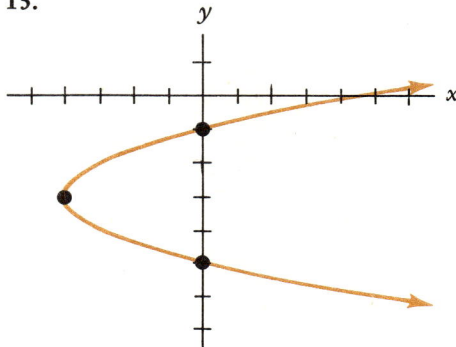

domain: $[-4, +\infty)$

range: $(-\infty, +\infty)$

not a function

ANSWERS TO ODD-NUMBERED EXERCISES

17.

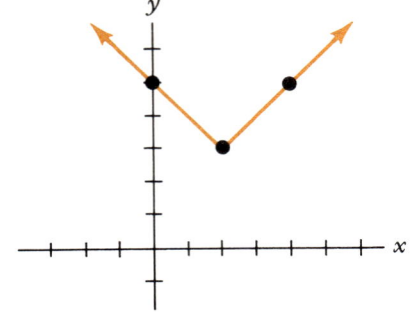

domain: $(-\infty, +\infty)$

range: $[3, +\infty)$

function

19.

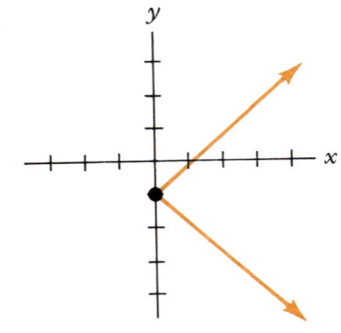

domain: $[0, +\infty)$

range: $(-\infty, +\infty)$

not a function

21. $\dfrac{(x+3)^2}{9} + \dfrac{(y-5)^2}{25} = 1$, vertical ellipse

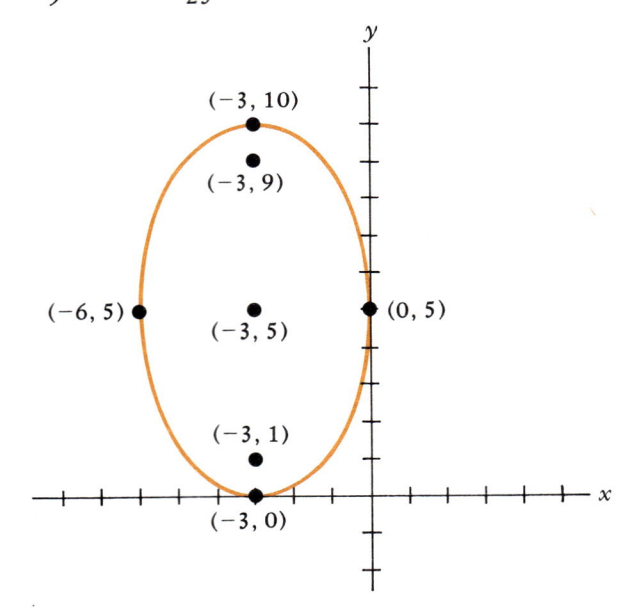

23. $\dfrac{(x-2)^2}{25} - \dfrac{(y+1)^2}{16} = -1$, vertical hyperbola

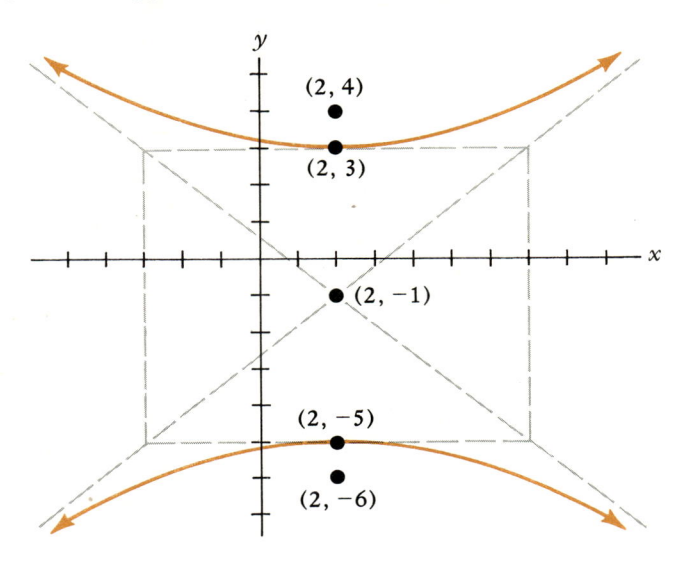

25. The single point $(-3, 4)$. This is often referred to as a *degenerate circle*.
27. $3x + 5y = -39$ 29. $5x - y = 3$ 31. $3x^2 + 3x - 1$, $(-\infty, +\infty)$
33. $3x^3 - x^2 - 2x$, $(-\infty, +\infty)$ 35. $3x^2 - 4x + 1$, $(-\infty, +\infty)$
37. $\dfrac{-2-x}{x(x+1)}$, $(-\infty, -1) \cup (-1, 0) \cup (0, +\infty)$
39. $\dfrac{x}{2(x+1)}$, $(-\infty, -1) \cup (-1, 0) \cup (0, +\infty)$
41. $(f \circ g)(x) = x$, $(g \circ f)(x) = x$ 43. $f^{-1}(x) = -\sqrt{\dfrac{x+4}{3}}$, $x \geq -4$
45. $f^{-1}(x) = \dfrac{1-3x}{x}$, $0 < x < \dfrac{1}{3}$

Chapter 3 Practice Test (page 107)

1. a. $D: \left[\dfrac{5}{2}, +\infty\right)$, $R: [0, +\infty)$, yes   b. $D: [-3, 3]$, $R: [-3, 3]$, no
   c. $D: (-\infty, +\infty)$, $R: [0, +\infty)$, yes
3. $(x-5)^2 + (y+2)^2 = 36$, center: $(5, -2)$, radius: 6

5. $\dfrac{(x-2)^2}{4} - \dfrac{(y+4)^2}{16} = 1$, horizontal hyperbola

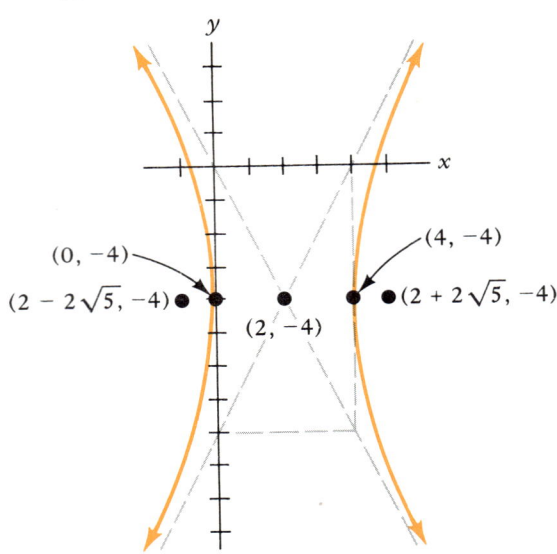

7. a. $\dfrac{6x^2 - 11x + 4}{2x - 3}$, $\left(-\infty, \dfrac{3}{2}\right) \cup \left(\dfrac{3}{2}, +\infty\right)$

b. $\dfrac{1}{6x^2 - 11x + 3}$, $\left(-\infty, \dfrac{1}{3}\right) \cup \left(\dfrac{1}{3}, \dfrac{3}{2}\right) \cup \left(\dfrac{3}{2}, +\infty\right)$

c. $\dfrac{1}{6x - 5}$, $\left(-\infty, \dfrac{5}{6}\right) \cup \left(\dfrac{5}{6}, +\infty\right)$

CHAPTER 4

4.1.1 (page 113)

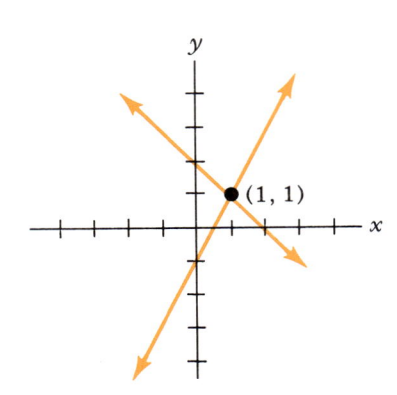

1.

3.

5.

7.

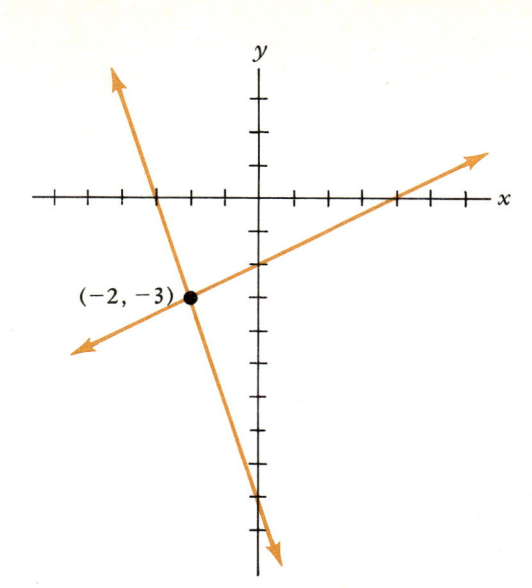

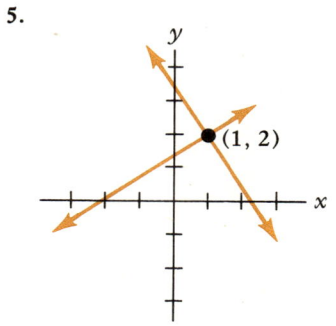

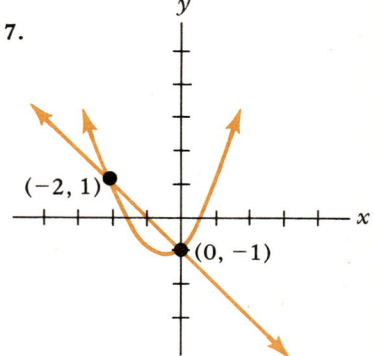

9. one point, no points, infinitely many points

4.2.1 (page 116)
1. (1, 2)   3. (0, 5)   5. (2, −1)   7. (4, 9)

4.2.2 (page 118)
1. (2, −3)   3. (1, −1)   5. (2, 0)   7. (−2, 6)

4.2.3 (page 120)
1. (2, −3)   3. (1, 4)   5. (0, −2)   7. dependent   9. (−3, 7)
11. $\left(\frac{1}{2}, 3\right)$ 13. dependent   15. (3, −8)

17. plane (500 mph), wind (100 mph)   19. boat (12 mph), current (3 mph)

4.3.1 (page 126)
1. $(1, 2, 3)$ 3. $(-2, 1, 3)$ 5. $(2, 1, -3)$ 7. $(-2, 1, 0)$
9. 5 nickels, 10 dimes, 13 quarters
11. 2 liters of 5%, 3 liters of 20%, 4 liters of 50%

4.4.1 (page 130)
1. $(4, -1)$ 3. $(3, -9)$ 5. $(2, 5, 0)$ 7. $\left(\frac{1}{2}, -1, 5\right)$
9. $10,000 at 5%, $15,000 at 8%

4.5.1 (page 132)
1. 7   3. -3 5. 11   7. -2 9. 0   11. -101 13. 0   15. 54

4.5.2 (page 135)
1. -9 3. 3   5. 81   7. 46   9. 0

4.6.1 (page 140)
1. $(1, 2)$ 3. $(2, 3)$ 5. dependent   7. $(3, 0)$ 9. $(5, -1)$
11. $(-2, 1)$ 13. $(4, -1)$ 15. $(-1, 5)$

4.7.1 (page 142)
1. $(1, 2, 3)$ 3. $(2, 3, -5)$ 5. $(-3, 1, 0)$ 7. $(1, -2, 4)$ 9. $(3, -10, 7)$

4.8.1 (page 145)
1.   3.

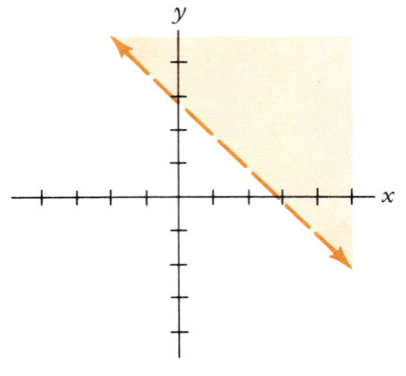

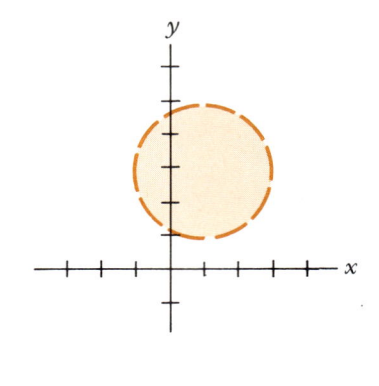

5.

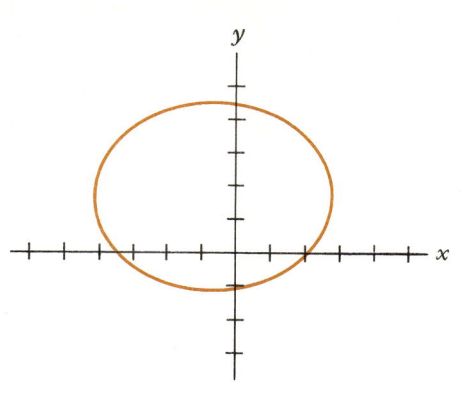

7.

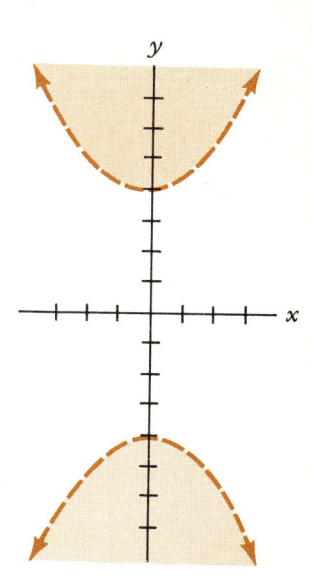

4.8.2 (page 149)

1.

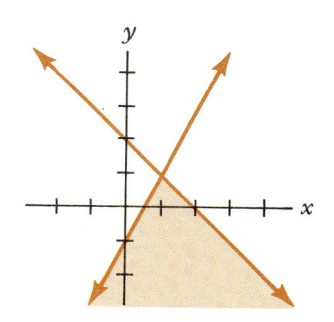

3.

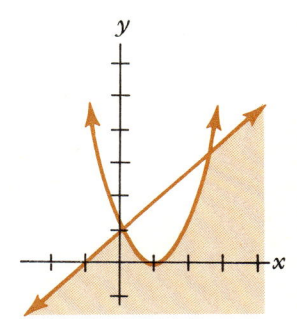

5.

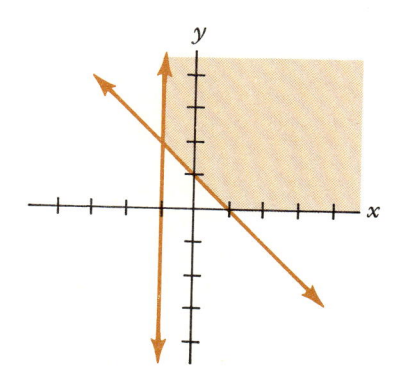

7.

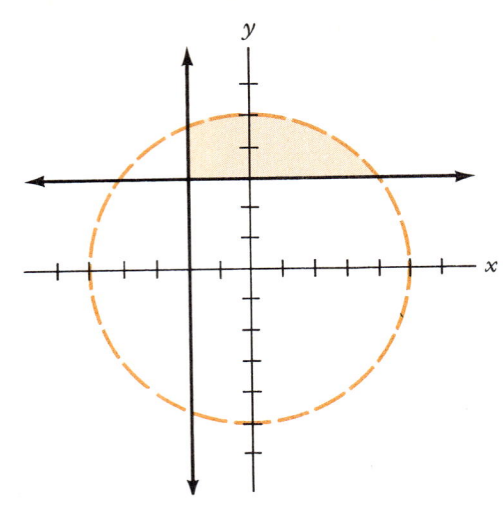

9. minimum $(10, -20)$, maximum $\left(\dfrac{45}{2}, 5\right)$

Chapter 4 Review (page 150)

1. $(1, -2)$

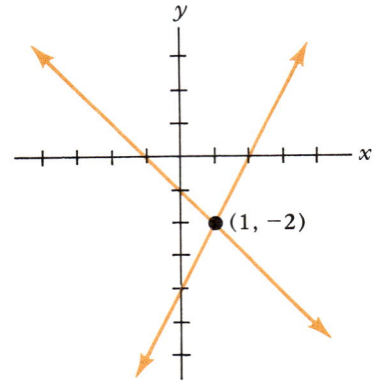

3. $(-2, 1)$

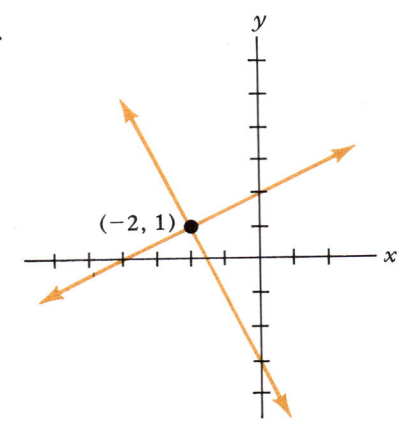

5. $(-1, 0)$, $(0, -1)$

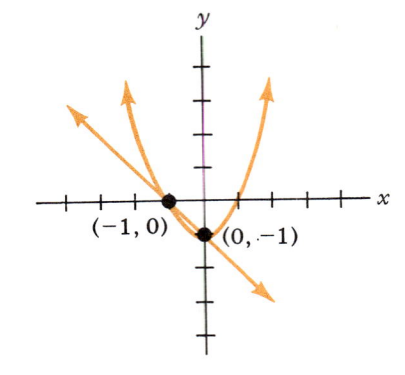

7. inconsistent   9. dependent   11. $(1, 2, -1)$ 13. $(1, 0, 3)$
15. $(-3, 1, -2)$ 17. $(0, 2)$ 19. $(0, 0, 4)$ 21. 12   23. 0   25. 36
27. $(1, -5)$ 29. $(0, -3)$ 31. $(1, 2, -1)$ 33. $(1, 0, 3)$ 35. $(-3, 1, -2)$
37. 5

39.

41.

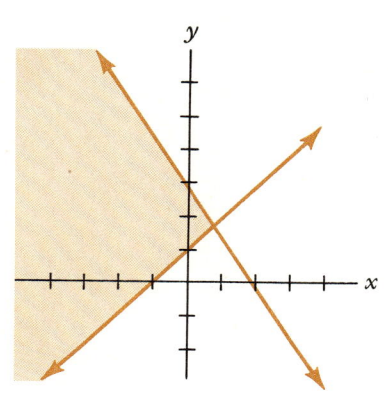

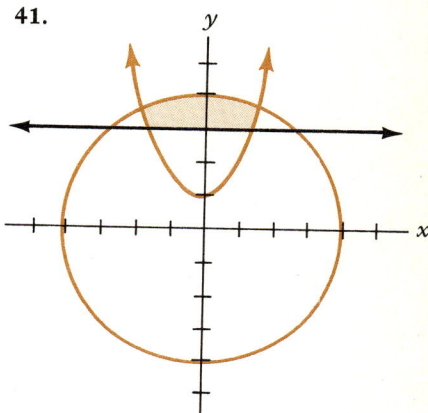

43. $8,000 in bond A, $4,000 in bond B

Chapter 4 Practice Test (page 152)

1.

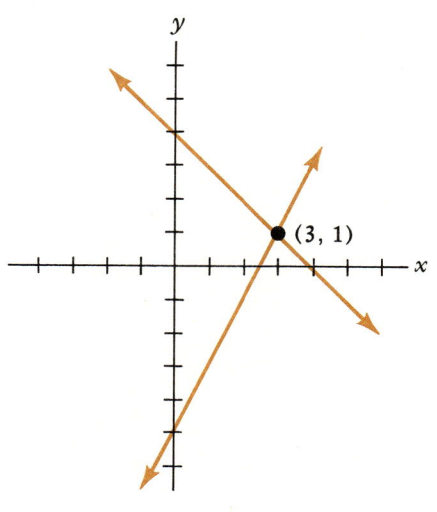

3. $(3, -2, 6)$ 5. $(5, -4)$

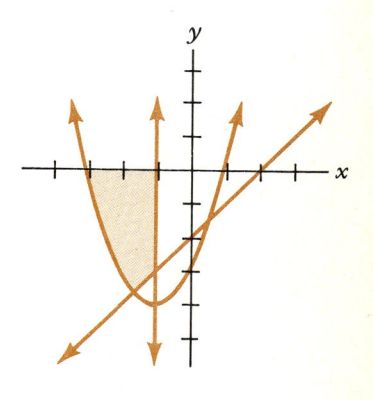

ANSWERS TO ODD-NUMBERED EXERCISES

CHAPTER 5

5.1.1 (page 158)
 1. $\log_3 9 = 2$ 3. $\log_5 125 = 3$ 5. $\log_3 27 = 3$ 7. $\log_2 \frac{1}{8} = -3$
 9. $\log_2 \frac{1}{16} = -4$ 11. $2^2 = 4$ 13. $4^2 = 16$ 15. $3^4 = 81$
 17. $5^3 = 125$ 19. 81   21. 2   23. 5   25. $\frac{1}{9}$ 27. 49   29. 1
 31. 3   33. -2 35.

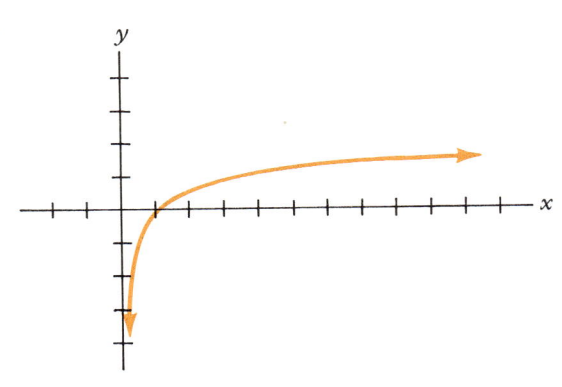

5.2.1 (page 161)
 1. $2y$ 3. $2x + y$ 5. $x - 2y$ 7. $2y - 2x$ 9. 2   11. 4   13. -2
 15. 5   17. $0.5514 + 2$ 19. $0.5514 - 2$ 21. $0.5514 - 3$

5.3.1 (page 162)
 1. 230   3. 0.00082   5. 719,000   7. 5.37   9. 0.0000326

5.3.2 (page 163)
 1. 4.6×10^1 3. 5.28×10^3 5. 1.26×10^{-1} 7. 1.02×10^4
 9. 6.19×10^0 11. 3.01×10^{-2}

5.3.3 (page 165)
 1. 0.3711   3. 0.2856   5. 0.8976   7. $0.5694 + 1$ 9. $0.5172 + 2$
 11. $0.5145 + 4$ 13. $0.1461 - 2$ 15. $0.4900 + 5$ 17. $0.3284 - 3$
 19. $0.4771 - 3$ 21. 3.5229

5.4.1 (page 168)
 1. 4.09   3. 741   5. 7,580   7. 0.045

5.4.2 (page 170)
1. 0.4343   3. 1.8004   5. 3.9676   7. 0.4751 − 1
9. Interpolation treats the function between two points as linear. The logarithmic function and the squaring function are not linear.

5.4.3 (page 172)
1. 5.315   3. 0.01716   5. 10,870   7. 0.002086

5.5.1 (page 174)
1. 11.96   3. 1,334   5. 1.692   7. 6.696   9. 5.244

5.6.1 (page 176)
1. 9.936   3. 0.001945   5. 0.000000001175   7. 16   9. 0.3084

5.7.1 (page 178)
1. 4.858   3. 2.921   5. 2.101   7. 3.914

5.8.1 (page 182)
1. 2   3. no solution   5. $\frac{1}{9}$ 7. no solution   9. 3,4   11. 4
13. $\frac{\log 7}{\log 2}$ 15. $\frac{\log 9 - \log 16}{\log 8}$ 17. $\frac{5 \log 3 - \log 2}{\log 2 - 2 \log 3}$
19. $\frac{\log 5 - \log 3}{\log 3 + 2 \log 5}$ 21. $\frac{1,006}{3}$ 23. 9 years   25. 7,601 years

Chapter 5 Review (page 183)
1. 64   3. 9   5. −3   7. $3x$ 9. $3x + 2y$ 11. 0.9053
13. 4.6902   15. 0.1931 − 4   17. 908   19. 0.6539 − 1   21. 21.87
23. 0.08112   25. 0.1293   27. 0.0001406   29. 4.149   31. 3.351
33. 0.5092   35. −0.191   37. 11   39. $\frac{\log 7 - \log 3}{\log 3}$ 41. 14 years

Chapter 5 Practice Test (page 184)
1. $\log_3 9 = 2$
3. a. 2   b. 9   c. 3   d. −4   e. 5
5. a. 1.3927   b. 0.1931 − 4   c. 0.8539   d. 4.1519
7. a. 13,140   b. 23.39

ANSWERS TO ODD-NUMBERED EXERCISES

CHAPTER 6

6.1.1 (page 191)
1. a. -18 b. 0 c. 12 d. 0 e. 0 f. 0
3. no   5. $J(x) = 0$ 7. yes, the set of polynomials of degree zero
9. a. 3 b. 6 c. 6 d. 9

6.2.1 (page 197)
1. a. $3x^2 - 4x + 8; -10$ b. $x^3 + 5x^2 - 2; -3$
   c. $3x^3 + 2x; 5x + 1$ d. $x^4 - 4x^3 + 3x^2 - x + 4; -1$
   e. $x^3 - x^2 + x + 1; -2x + 9$ f. $6x^3 + 18x^2 + 3x + 9; 0$
   g. $2x^2 + 3x - 2; 0$
3. no   5. yes   7. yes

6.3.1 (page 201)
1. $x^2 - x - 1, r = 1$ 3. $2x^3 - 3x^2 + 6x - 12, r = 23$
5. $x^5 + 2x^2 - x, r = 1$ 7. -9 9. 11 11. no   13. yes   15. yes

6.4.1 (page 204)
1. a. 2 b. 2 c. 3 d. 5 e. 1 f. 4
3. $x^3 - (1 + 4\sqrt{2})x^2 + (6 + 4\sqrt{2})x - 6$ 5. $x^3 - 2ix^2 + x - 2i$
7. $x^4 - 29x^2 + 100$ 9. $-5, 6$ 11. $\pm 1, \pm i$ 13. ± 6 15. $\pm i$
17. $\left(\dfrac{-1 \pm i\sqrt{3}}{2}\right), 1$

6.5.1 (page 207)
1. a. 1 b. 3 or 1 c. no, $f(0) = -4$ 3. The constant term must be zero.
5. $f(x)$ has one sign variation; $f(-x)$ has one sign variation.
7. $f(x)$ and $f(-x)$ have no sign variations and $f(0) \neq 0$.   9. 2

6.6.1 (page 212)
1. $\pm 1, 2$ 3. $-2, \dfrac{1}{2}, 3$ 5. $-\dfrac{1}{2}, \dfrac{1}{3}, \dfrac{2}{3}$ 7. $-\dfrac{3}{2}, -1 \pm \sqrt{3}$
9. $\pm 1, -3, \dfrac{1}{2}$ 11. $-1, \dfrac{2}{3}, 2 \pm i$ 13. $5, 8, 11, \dfrac{-1 \pm i\sqrt{3}}{2}$

15. If $a_0 = 1$, then $q = 1$ since the only exact divisor of 1 is 1. Thus $\dfrac{p}{q} = \dfrac{p}{1} = p$ (an integer).

6.7.1 (page 217)
1. $(-2, -1), (-1, 0), (1, 2)$ 3. $(-7, -6), (0, 1), (5, 6)$
5. Either three real zeros in $(1, 2)$, or one real in $(1, 2)$ and two nonreals.
7. 2.06 9. -1.54 11. $-1.17, -0.36, 3.53$

6.8.1 (page 222)

1. $\dfrac{1}{3(x-2)} - \dfrac{1}{3(x+1)}$ 3. $\dfrac{1}{x-2} - \dfrac{2}{3(x-1)} + \dfrac{1}{3(2x+1)}$

5. $-\dfrac{1}{x} + \dfrac{2}{x+1} - \dfrac{1}{(x+1)^2} + \dfrac{2}{(x+1)^3}$ 7. $-\dfrac{3}{x} + \dfrac{3}{x-1} + \dfrac{1}{x+1}$

9. $\dfrac{1}{x-1} + \dfrac{2x+3}{x^2+1}$ 11. $\dfrac{1}{x+1} + \dfrac{2x-2}{x^2-x+3}$

13. $\dfrac{x+3}{x^2+1} - \dfrac{x}{x^2+2}$ 15. $\dfrac{1}{x+3} - \dfrac{x-3}{x^2+2} - \dfrac{5x}{(x^2+2)^2}$

Chapter 6 Review (page 223)

1. 5   3. 3   5. $a=5, b=-1, c=11$ 7. -404 9. -70 11. yes
13. $(x+6)(x^4 - 6x^3 + 6x^2 + 4x - 24) + 41$ 15. 15   17. no   19. 6
21. $x^3 - (3-\sqrt{5})x^2 - (10+3\sqrt{5})x + 30$ 23. $-1, \dfrac{1 \pm i\sqrt{3}}{2}$ 25. 2 or 0
27. 0   29. 1   31. $-1, 3, 5$ 33. 1 (multiplicity 2), 3 (multiplicity 2)
35. $-1, \dfrac{2}{3}, 4, \pm \dfrac{i}{2}$ 37. $(-1,0), (2,3), (4,5), (9,10)$ 39. -2.66
41. $\dfrac{5}{x-2} - \dfrac{2}{x+5}$ 43. $\dfrac{2x+1}{x^2+2x+4} + \dfrac{6}{x-2}$
45. $\dfrac{2x-3}{x^2+2x+3} - \dfrac{3x+1}{(x^2+2x+3)^2}$

Chapter 6 Practice Test (page 225)

1. 7   3. $f(-5)=0$ 5. $x^3 + 2x^2 - 5x - 6$ 7. $\left\{\pm 1, \pm 5, \pm \dfrac{1}{3}, \pm \dfrac{5}{3}\right\}$

9. $\dfrac{1}{2}, \pm 2i$ 11. $(-5,-4), (-2,-1), (3,4)$

CHAPTER 7

7.2.1 (page 235)

1. $a_1 = 1, a_2 = 4, a_3 = 7, a_4 = 10, a_{10} = 28$
3. $a_1 = 4, a_2 = 7, a_3 = 12, a_4 = 19, a_{10} = 103$
5. $a_1 = 0, a_2 = 0.0753, a_3 = 0.0795, a_4 = 0.0753, a_{10} = 0.05$
7. $a_1 = -4, a_2 = -19, a_3 = -94, a_4 = -469$
9. $a_1 = 1, a_2 = 4, a_3 = 19, a_4 = 364$ 11. 50   13. 36   15. $\dfrac{119}{16}$

17. $\displaystyle\sum_{i=1}^{4} (i+1)^2$ 19. $\displaystyle\sum_{i=1}^{6} (-1)^{i+1}$

ANSWERS TO ODD-NUMBERED EXERCISES

7.3.1 (page 238)
1. 3, 8, 13, 18, 23, 28   **3.** 52   **5.** −7   **7.** 193   **9.** 200   **11.** −32
13. −15   **15.** 10   **17.** 5   **19.** 10   **21.** yes   **23.** $4,400

7.4.1 (page 242)
1. 2, 6, 18, 54, 162   **3.** 9,375   **5.** 384   **7.** 6   **9.** −3
11. log 81, log 6,561   **13.** 6, 12, 24, 48   **15.** 27 days

7.4.2 (page 243)
1. 8   **3.** sum does not exist   **5.** $4(2 + \sqrt{2})$ **7.** 2
9. a. $a_1 = 0.36, r = 0.01, s = \frac{4}{11}$ **b.** $a_1 = 0.345, r = 0.001, s = \frac{115}{333}$
11. 84 meters

Chapter 7 Review (page 244)
1–6. Proofs may vary.   **7.** $a_1 = -3, a_2 = 4, a_3 = 23, a_4 = 60, a_8 = 508$
9. $a_1 = 1, a_2 = 2, a_3 = 6, a_4 = 21$ **11.** $\frac{x^3}{4} + \frac{x^4}{8} + \frac{x^5}{12} + \frac{x^6}{16}$
13. $\sum_{i=1}^{6} (4i - 3)$ **15.** −6   **17.** 10,000   **19.** 0
21. $\sqrt{2}, 2, 2\sqrt{2}, 4$ **23.** $-\frac{1}{2}$ **25.** $-\frac{27}{14}, 0$ **27.** $\frac{2}{3}$

Chapter 7 Practice Test (page 245)
1. If $n = 1$, then
$$1(1 + 1) = 2$$
Assume the statement true for $n = k$, then the sum of the first k terms is $k(k + 1)$. For $n = (k + 1)$ we have
$$k(k + 1) + 2(k + 1) = (k + 1)(k + 2)$$
$$= k^2 + 3k + 2$$
$$= (k + 1)[(k + 1) + 1]$$
3. 48   **5.** $4\frac{1}{3}$ **7.** 10   **9.** 2, 6, 18 or −2, 6, −18

CHAPTER 8

8.1.1 (page 253)
1. 120   **3.** 144   **5.** 360   **7.** 30,240   **9.** 120   **11.** 40,320
13. 120   **15.** 7,893,600   **17.** 1,320   **19.** 250

8.2.1 (page 255)
1. 120 3. 495 5. $\binom{n}{r} = \frac{n!}{(n-r)!r!} = \binom{n}{n-r}$ 7. 10 9. 56
11. 462 13. 2,598,960 15. 1,024

8.3.1 (page 261)
1. $x^4 + 8x^3 + 24x^2 + 32x + 16$
3. $x^6 - 6x^5y + 15x^4y^2 - 20x^3y^3 + 15x^2y^4 - 6xy^5 + y^6$
5. $128a^7 + 224a^6b + 168a^5b^2 + 70a^4b^3 + \frac{35}{2}a^3b^4 + \frac{21}{8}a^2b^5 + \frac{7}{32}ab^6 + \frac{b^7}{128}$
7. $a^{100} - 300a^{99}b + 44,550a^{98}b^2 - 4,365,900a^{97}b^3$ 9. $-560a^8b^6$
11. 108,243,216 13. 1.172 15. 7.071 17. 1.005 19. 0.032

8.4.1 (page 264)
1. $\frac{1}{2}$ 3. a. $\frac{1}{52}$ b. $\frac{1}{13}$ c. $\frac{1}{4}$ d. $\frac{3}{13}$
5. a. $\frac{5}{31}$ b. $\frac{9}{31}$ c. $\frac{12}{31}$ d. $\frac{21}{31}$ e. $\frac{15}{31}$

8.4.2 (page 267)
1. $\frac{1}{36}$ 3. $\frac{1}{4}$ 5. $\frac{1}{2}$ 7. $\frac{2}{7}$ 9. $\frac{1}{1,320}$ 11. $\frac{1}{635,013,559,600}$

8.4.3 (page 270)
1. $\frac{18}{23}$ 3. $\frac{2}{13}$ 5. $\frac{1}{6}$ 7. $\frac{4}{13}$ 9. $\frac{13}{48}$

8.4.4 (page 272)
1. 1:5 3. 1:12 5. a. 1:35 b. 1:5 7. 1:220 9. 33:66,607

Chapter 8 Review (page 273)
1. 40,320 3. 720 5. 362,880 7. 604,800 9. 220 11. 127
13. $64x^6 + 576x^5y + 2160x^4y^2 + 4320x^3y^3 + 4860x^2y^4 + 2916xy^5 + 729y^6$
15. $3360y^{12}$ 17. 0.32768 19. 2.087 21. $\frac{1}{78}$ 23. $\frac{14}{31}$ 25. 1:31
27. 649,739:1

Chapter 8 Practice Test (page 274)
1. 12 3. 816 5. 210 7. 3.87 9. $\frac{7}{13}$

Index

Abscissa, 69
Absolute value, 5
Addition method, 116–118
Additive identity, 3
Additive inverse, 3
Algebraic expression, 16–20
Algebraic fractions, 24–26
Antilogarithm, 166
Approximation of zeros of
    polynomials, 217
Arithmetic mean, 239
Arithmetic sequence, 236–238
    definition, 236
    "nth" term, 236
    sum of, 237
Associative property
    of addition, 3
    of multiplication, 3
Asymptote, 87

Base of logarithm, 156
Binomial theorem, 256–261
    to find roots, 260–261
    to expand binomials, 257–260
Binomials, 19
Bounds for zeros of polynomials,
    212–217

Center
    of circle, 80
    of ellipse, 83
    of hyperbola, 87
Characteristic, 164
Circle, 79, 80–82
    center of, 80
    radii of, 80
    standard form, 80
Closed interval, 52
Coefficient, 19
Combinations, 254–255
    definition, 254
Common logarithms, 164

Commutative property
    of addition, 3
    of multiplication, 3
Completing the square, 40
Complex fractions, 25
Complex numbers, 4, 42–45
    addition of, 44
    conjugates, 45
    division of, 45
    inverse of, 45
    multiplication of, 44
    subtraction of, 44
Composite function, 99
Conic sections, 79–92
Conjugates, 45
Coordinate axes, 72
Coordinates, 72
Counting numbers, 2
Cramer's Rule, 137

Degree, 32
    of a polynomial, 189
Dependent equations, 118
Descartes' Rule of Signs, 206–207
Determinants, 136–142
Difference function, 98
Directrix, 90
Discriminant, 41
Distance formula, 73–74
Distributive property, 4
Division
    of polynomials, 191–197
    synthetic, 198–201
Division algorithm
    for integers, 192
    for polynomials, 194
Domain, 69

"e" (base of the natural logarithm),
    177
Elementary row operations, 128
Ellipse, 79, 82–85

    foci of, 82
    horizontal, 84
    standard form, 83
    vertical, 84
    vertices of, 84
Equation, 32
Equations
    dependent, 118
    equivalent, 32
    exponential, 180
    first-degree, 32–35
    higher-degree, 36
    inconsistent, 118
    independent, 118
    logarithmic, 179
    quadratic, 36–47
    quadratic in form, 48–51
    systems of, 110–125
Equivalent equations, 32
Expanding binomials using the
    binomial theorem, 257–260
Exponential equations, 180
Exponential form, 157
Exponents, 7–9
    fractional, 8
    laws of, 7–8
    zero, 8
Extraneous root, 50

Factor theorem, 195
Factorial notation, 252
Factoring, 21–23
Factors, 18
Field, 4
Finite sequence, 232
First-degree equations, 32–35
First-degree inequalities, 51–55
Fractional exponents, 8
Fractions, 2
    algebraic, 24–26
    complex, 25
    fundamental principle of, 24

Function, 69
Functions
  composite, 99
  difference, 98
  inverse, 100-104
  linear, 93-97
  logarithmic, 155
  product, 98
  quotient, 98
  sum, 98
Fundamental counting
    principle, 251
Fundamental principle of
    fractions, 24

Geometric means, 241
Geometric sequence
  definition, 240
  "nth" term of, 240
  sum of finite, 240
  sum of infinite, 243
Geometric sequences, 239-243
Graphs of relations, 71-78

Harmonic mean, 239
Harmonic progression, 239
Horizontal ellipse, 84
Horizontal hyperbola, 87
Horizontal parabola, 91
Hyperbola, 79, 87-89
  horizontal, 87
  standard form, 87
  vertical, 87

Identity element, 103
Image set, 69
Images, 69
Inconsistent equations, 118
Independent equations, 118
Independent events, 262, 265
Index number, 10
Inequalities, 51-61
  first-degree, 51-55
  graphical method of solving, 56-58
  systems of, 143-149
Inequality, 32
  definition, 5
Infinite sequence, 232
Integers, 2
Integral domain, 191
Interpolation, 168
Intersection symbol, 98
Interval
  closed, 52
  half-closed, 52
  half-open, 52
  open, 52
Inverse
  of a complex number, 45
  of a function, 103

Inverse function, 100-104
Inverse relation, 101
Irrational number, 3

Laws of exponents, 7-8
Laws of logarithms, 159-160
Like terms, 19
Line segment, length of, 73
Linear equations, standard form, 136
Linear functions, 93-97
Linear programming, 143
Literal factors, 19
Logarithmic equations, 179
Logarithmic form, 157
Logarithmic function, 155
Logarithms, 156-182
  base of, 156
  definition, 156
  to different bases, 176-178
  laws of, 159-160
  table of, 276-277
  use of table of, 161-165
Lower bounds for zeros of
    polynomials, 213

Mantissa, 164
Mathematical induction, 228-231
Matrices, 126-130
Matrix
  augmented, 127
  of coefficients, 127
  cofactor, 133
  determinant, 131, 133
  main diagonal, 128
  minor, 132
  triangular form, 128
Matrix method, 129
Midpoint formula, 74
Monomial, 18
Multiplicative identity, 3
Multiplicative inverse, 3
Mutually exclusive events, 268

Natural logarithms, 178
Natural number "e", 177
Negative of a number, 3
Numbers
  counting, 2
  irrational, 3
  rational, 2
  real, 2

Odds, 271
Open interval, 52
Ordered pairs, 68, 69, 72
Ordered triple, 121
Ordering property, 4
Ordinate, 69
Origin, 72

Parabola, 76, 90-92
  directrix of, 90
  focus of, 90
  horizontal, 91
  vertical, 91
Partial fractions, 218-222
Pascal's triangle, 259-260
Permutations, 250-253
  definition, 251
Point-slope form, 95
Polynomial, 18
Polynomials
  degree of, 189
  division of, 191-197
  properties of, 188-189
  real, 189
  zeros of, 189
Positive numbers, 4
Power, 7
Principle of mathematical
    induction, 228
Principle root, 10-11
Probability, 262-272
  of mutually exclusive events,
      269-270
  of non-mutually exclusive
      events, 270
  of successive events, 267
  of two independent events, 265
Product function, 98
Properties
  of polynomials, 188-189
  of real numbers, 3

Quadrants, 72
Quadratic equations, 36-47
  complex roots of, 46-47
  nature of roots of, 41
  standard form, 37
Quadratic formula, 39
Quotient function, 98

Radical sign, 10
Radicals, 10-15
  simplification of, 12-15
Radicand, 10
Range, 69
Rational numbers, 2
Real numbers, 2-6
Real polynomials, 189
Reciprocal, 3
Relation, 69
Remainder theorem, 194

Scientific notation, 162
Sequences, 231-235
  arithmetic, 236-238
  definition, 232
  finite, 232

geometric, 239–243
  infinite, 232
Series
  definition, 234
  sigma notation for, 234
Sigma notation, 234
Sign variation, 206
Slope, 93–94
Slope-intercept form, 95–96
Solution set, 32
Standard form
  of circle, 80
  of ellipse, 83
  of hyperbola, 87
  of parabola, 91
  of quadratic equation, 37
  of straight line, 93
  of system of two linear
    equations, 136
Straight line
  point-slope form of, 95
  slope of, 93

slope-intercept form of, 95–96
standard form, 93
two-point form of, 94
Substitution method, 113–116
Sum
  of arithmetic series, 237
  of finite geometric sequence, 240
  of infinite geometric sequence, 243
Sum function, 98
Synthetic division, 198–201
Systems of equations
  algebraic solution of, 113–125
  graphical solution of, 110, 112
Systems of inequalities, 143–149

Table of logarithms, use of, 161–165
Term, 18
Trinomials, 19
Two-point form, 94

Upper bounds for zeros of
  polynomials, 213

Value of an algebraic expression, 17
Variables, 17
Verbal problems, 59–61
Vertical ellipse, 84
Vertical hyperbola, 87
Vertical parabola, 91

x-axis, 72

y-axis, 72
y-intercept, 95

Zero exponent, 8
Zero of a polynomial, 189
Zeros of polynomials
  approximating, 212–217
  with complex coefficients, 202–204
  with integral coefficients,
    208–211
  with rational coefficients, 208–211
  with real coefficients, 205–207

INDEX

FORMULAS

FACTORING

$x^2 - y^2 = (x - y)(x + y)$

$x^3 - y^3 = (x - y)(x^2 + xy + y^2)$

$x^3 + y^3 = (x + y)(x^2 - xy + y^2)$

$(x + y)^2 = x^2 + 2xy + y^2$

QUADRATIC FORMULA

If $ax^2 + bx + c = 0$, then

$$x = \frac{-b \pm \sqrt{b^2 - 4ac}}{2a}$$

SEQUENCES & SERIES

nth term of an arithmetic sequence

$$a_n = a + (n - 1)d$$

Sum of the first n terms of an arithmetic series

$$S_n = \frac{n}{2}(a_1 + a_n)$$

nth term of a geometric sequence

$$a_n = a_1 r^{n-1}$$

Sum of the first n terms of a geometric series

$$S_n = \frac{a_1 - a_1 r^n}{1 - r}$$

Sum of an infinite geometric series

$$S = \frac{a_1}{1 - r} \qquad |r| < 1$$